Chicken

Environmental History

Adrian Howkins, *The Polar Regions*
Jon Mathieu, *The Alps*

Chicken

A History from Farmyard to Factory

Paul R. Josephson

polity

First published in 2020 by Polity Press

Polity Press
65 Bridge Street
Cambridge CB2 1UR, UK

Polity Press
101 Station Landing
Suite 300
Medford, MA 02155, USA

ISBN-13: 978-1-5095-2591-1

A catalogue record for this book is available from the British Library.

Library of Congress Cataloging-in-Publication Data

Names: Josephson, Paul R., author.
Title: Chicken : a history from farmyard to factory / Paul R. Josephson.
Description: Cambridge, UK ; Medford, MA : Polity Press, 2020. | Includes bibliographical references and index. | Summary: "The irresistible rise and fall of the world's favourite farm animal"-- Provided by publisher.
Identifiers: LCCN 2019045569 (print) | LCCN 2019045570 (ebook) | ISBN 9781509525911 (hardback) | ISBN 9781509525942 (epub)
Subjects: LCSH: Broilers (Chickens)--History. | Factory farms.
Classification: LCC SF498.7 .J67 2020 (print) | LCC SF498.7 (ebook) | DDC 636.5--dc23
LC record available at https://lccn.loc.gov/2019045569
LC ebook record available at https://lccn.loc.gov/2019045570

Typeset in 10.75 on 14 Adobe Janson by
Servis Filmsetting Ltd, Stockport, Cheshire
Printed and bound in Great Britain by CPI Group (UK) Ltd, Croydon

For further information on Polity, visit our website:
politybooks.com

Contents

For Isaac, Nina and Emile

They bring smiles to the table,
but only one is a vegetarian.

Acknowledgments

A number of people helped in writing this book, and thank you to all of them. Colby College provided financial support for *Chicken* with a travel grant, and my students in the lecture "Luddite Rantings" responded to my lectures on broilers with well-seasoned questions. I presented early thoughts on chicken at the "Consuming the World: Eating and Drinking in Culture, History, and Environment" workshop at the Rachel Carson Center for Environment and Society in Munich, Germany, in March 2016, and thank the organizers Michelle Mart, Dan Philippon and Hanna Schösler for inviting me to participate and learn about food from them and the other participants. Deborah Fitzgerald read the draft of the paper from the conference and shared her broad knowledge on agriculture, technology and industry. My editor at Polity, Pascal Porcheron, has been supportive during the entire process; his in-depth comments and suggestions make this book stronger. The two anonymous reviewers offered excellent suggestions on how to raise, shape, cut and bread my writing. Indiana Jones, Lego-master and friend, again assembled a superior index. Willie "Pops" Stargell and Roberto Clemente, as always, provided inspiration to move ahead when the hunting and pecking got tough. Pops was especially instrumental in motivating this book. He owned a chicken restaurant in Pittsburgh's Hill District, and any customer in the restaurant got free chicken when

Stargell hit a home run. Bob Prince, the Pirate radio announcer, came to announce Stargell home runs with the yell "Chicken on the Hill with Will." Willie took me to a baseball game when I was 13 at Forbes Field. Thank you, Willie. Tatiana Kasperski ate chicken with me when I prepared it, cooked chicken for me, added it to *kholodnik*, listened to my ideas and plucked them carefully. Thank you, Tanya.

The publishers would like to thank the following for permission to reproduce the images:
Frontispiece: Gamborg Gallery, Moscow, Russia; figure 1, bariskarad eniz/istock; 2, Nastasic/istock; 3, Alf Ribeiro/Shutterstock; 4, travelview/ istock; 5: Carol M. Highsmith Archive, Library of Congress, Prints and Photographs Division; 6, zilli/ istock; 7, Africa Studio/ Shutterstock; 8, bluebird13/ istock; 9: Gloszilla Studio / shutterstock; 10, N-sky/ istock; 11, Photoagriculture/ shutterstock; 12, : ben/ Flickr (https://www.flickr.com/photos/16693144@N00/2512789670); 13, https://www.tvc.ru/news/show/id/57784#gid=gid_57784_0&pid=142286; 14, M M (Padmanaba01)/ Wikimedia Commons (https://commons.wikimedia.org/wiki/File:Chicken_market_in_Xining,_Qinghai_province,_China.jpg); 15, TommL/ istock.

"Ryaba, the Hen," Maria Uspenskaya (1987).

1 *Over the course of less than a century, chickens were transformed from farmyard birds to factory birds that are confined to sheds for all of their short, seven-week lives.*

Introduction: Egg First

> I must have been hungry,
> I ate another chicken.
> With my hands
> and noticed at the chicken dinner,
> that I had eaten a
> Cold and dead chicken.
>
> – Gunter Grass, "Saturn"[1]

Sometime in the early 2000s, when traveling abroad, I noticed a different taste in the fresh chicken distributed in the European Union (EU) from that delivered to the supermarket in the United States. One could say that US poultry left a bad taste in my mouth. Equally, one could say that there's no accounting for taste. To answer my taste confusion, I commenced investigation of how meat chickens are produced and delivered to stores. That olfactory and gustatory journey resulted in this political, cultural and environmental history of the broiler – precisely, of the meat chicken, an industrial object, hatched, shipped as chicks, raised to slaughter weight in but six or seven weeks, and then dispatched on an evisceration line for processing, transport and consumption around the globe. This study shows that the modern chicken, engineered over a century for rapid growth with meatier breasts and thighs, and the modern chicken industry, deployed for speed, efficiency and profit, leaves much to be desired from animal welfare, social and environmental perspectives. The broiler not only symbolizes, but actually is, billions of tons of chickens produced in tightly packed sheds, and billions of tons of poorly handled or stored fecal matter and offal including guts, feathers, and piles – millions – of dead animals. The broiler is generally produced by poorly paid contract workers at massive factory farms. On top of all this, the broiler presents a series of new health and safety risks, including new disease vectors for bacterial diseases and Avian Influenza.

The major production unit for the broiler, the concentrated animal feed operation (CAFO), consists across the globe of thousands of huge sheds erected in the last 10, 20, 30 years or so, with some of the sheds containing tens of thousands of fowl, each bird unable to move a more than a step, and each denied fresh air and free range – a kind of chicken gulag, where the purpose of each inmate is to devote all of its energy to the Chicken State before its eventual slaughter. The owners of CAFOs and other factory farms resist change in their chicken manufacture practices that might alter the production system in the direction of animal welfare or greater pollution control by pointing to the uncertainties and costs associated with change. Their owners claim the CAFOs need not be regulated more carefully for animal welfare because they meet national – and occasionally international – standards, they satisfy real and growing consumer demand for meat protein, and the solution to pollution and waste is being engaged head-on. Yet factory farms persist in pushing their employees, the environment and the chicken itself harder and harder, and when a factory farm is built in a pastoral, rural community, no one is happy, least of all the local residents who find their daily lives disrupted by the smell of ammonia, the sounds of trucks, and the domination of the local economy by an industry – the factory farm – that owes allegiance to owners, bosses and managers, not those residents. And, yes, European CAFO-generated chicken leaves a better taste in the mouth than the American broiler that suffers through weaker regulations in comparison to EU ones that cannot guarantee bird health or welfare.

Factory-farm chickens have spread across the world, and they are grown to maturity so quickly that it is difficult to determine even roughly what their total number is. I have tried to make sense of broilers' numbers. Over 53 billion broiler chickens are killed annually for their meat. The broiler's life is short and under the total control of its industrial handlers. Chickens peck out of eggshells after 21 days in an incubator. The incubator and the brooder, both nineteenth-century inventions, enabled separation of chicks from mother hens, and eggs from meat production. Being quite successful in producing eggs and freeing up human labor, the incubator and brooder accelerated the path to the complete industrialization of an animal. But, if chicks can walk at birth, then they are denied that possibility and stuffed into massive sheds. Were they only to be comfortable in their temporary homes with temperatures of 32 to 35 °C and humidity of 60 to 70 percent! But their purpose is not to move, but to eat and gain weight as quickly as possible – in the United States to, on average at 47 days, 2.6 kg, and in the European Union, at 42 days, to 2.5 kg. Since they reach slaughter weight within several weeks, they have poorly developed immune systems. Yet the overpacked sheds, filled with shit, feathers, shavings and dead birds, are breeding grounds for infections. The broilers must be fed or sprayed with antimicrobials ("vaccinated") against *Salmonella*, Newcastle disease virus, infectious bronchitis, Avian Influenzas, Marek's disease and others. The broiler is a bird, but it is also an industrial object, an output made of energy, chemical, and water input in carefully controlled environments.

The broiler is a prisoner in a technological panopticon with no prospects of hunting, pecking and roosting as chickens normally hunt, peck and roost. In the twentieth century, a kind of free-range life persists in many places, and even in urban and suburban backyard settings where they are pets, egg producers, insect swallowers, and also a source of food. In the unforgiving factory farm, various technologies of control focus the chicken's metabolism on rapid weight gain. But free-range chickens are in the distinct minority. At factory farms, chickens occasionally get natural daylight and natural ventilation, but they are rare exceptions. Only a few countries – the United Kingdom, the Netherlands and Sweden, for example – require windows in chicken sheds. The barren, seemingly infinite sheds that prevail in factory farms have only feeding and drinking points. Higher welfare would necessitate more space (reduced stocking density), slower-growing breeds, a later slaughter time, and access to outdoors. Very few chickens in the world (less than 1 percent) are raised free-range – that means at least half the time outdoors.[2] Since chickens are outdoors animals, this means a forced change in their behavior. They have been invited inside, under lock and key, for less than a century, but in this century they have become a major source of protein for billions of consumers.

This books aims at a global history of the industrial chicken. By the late 1950s, led by the United States, agricultural manufacturers had embarked on integrating the industry from bottom to top, from egg layers to chick producers to contractors responsible for shed raising, to processing plants for slaughter and reassembly as whole birds, parts, and various other products. The integrators own and control the delivery of the inputs – the birds, the feed, the antibiotics – all of which are to be tended and applied as specified by contract laborers or hourly paid workers. By the end of the century, this system had spread across the globe, with Brazil, China, France and the US among the leaders. These countries, often following the example of US factory-farm practices and businesses, accelerated a striking industrial transformation in agriculture. They have united nature and technology in a production paradigm. They have developed a single species of animal – moreover, a monoculture of animal – that is engineered to ensure a uniform product, the broiler. The broiler has been subjugated to the production ethos and the profit motive of the capitalist system, in this case primarily in the CAFO. The CAFO is a powerful tool: almost universally, nations have responded in a welcoming fashion to the broiler imperatives of efficiency, speed and industry concentration because people want cheap animal meat, and they have been slow to respond to growing community turmoil, pollution and public health threats.

Consumers and Their Role in Making the Modern-Day Broiler

A major reason for the success of the broiler has been growing consumer demand for chicken meat. Consumers are major actors in the chicken story – and in animal meat industries generally. Granted, in this book I focus on the production and technology sides of broilers. But it must be pointed out – if it is not obvious – that consumers north and south, east and west, in post-industrial nations and traditional peasant societies, have a growing, almost insatiable, appetite for meat. They want it on the table at home, and they want more poultry. They love it at KFC in China – the average Chinese person now eats more meat annually than the average American person. They want their nuggets when they pull up to a drive-in window of a fast-food joint in the US. France has become the fourth-largest producer and consumer of chicken in the world, satisfying to the French palate, and through exports also those of Saudis, South Africans, Spaniards and Brits. Americans consumed, in one day, 1.3 billion chicken wings during the broadcast of the 2018 football Super Bowl alone. July 29 in the US is National Chicken Wing Day. (Boneless wings, increasingly promoted by restaurants, are not wings at all, but slices of breast meat deep-fried like wings and served with sauces.)

Consumers seem content not to think about factory farms or the agricultural laborers who toil in them. They have offered little criticism – until recently – of deforestation of vast tracts of land to facilitate meat, especially beef, production that has finally captured attention owing to a summer 2019 crisis in Amazonia. And few people have embraced vegetarianism among the major meat-producing countries and regions – the EU, China, the US and Brazil. People want their chicken, and they want it now. They often care little about, or are unaware of, the additional costs to the globe of factory production of meat. Or perhaps they have become inured to factory-farm food because what they see in advertisements is gorgeously contrived steaming hot food to gobble, and what they see in stores comes in neat, clean, convenient packages. Over 7 billion people demand protein, and they eat a lot of chicken to get it. They want their chicken breasts breaded, fried, with parmesan, kung pao; their legs and thighs baked; their strips lightly floured, and so on – just as they enjoy eggs over-easy, sunny side up, coddled and poached. In many places, consumers have embraced foodie culture that has resulted in the glorification of oral gratification, but that some people believe provides moral cover to gluttony, and here, too, chicken is a presence.

Chicken has grown in demand and consumption for a variety of other reasons. In the 1970s and 1980s, doctors recommended that their patients eat less meat with saturated fat; the knowledge for patients that a diet high in fatty foods contributed to heart disease, arterial sclerosis and cancer had been available decades earlier.[3] But, after Senator

George McGovern's Select Committee on Nutrition and Human Needs published the *Dietary Goals for the United States* in 1977, many more physicians and citizens took notice. The National Academy of Sciences followed this in 1982 with the publication of *Diet, Nutrition, and Cancer*. Chicken provided an alternative safer and less fatty than beef and pork, and chicken production businesses took advantage of this situation in advertising campaigns. Some people argue that, as a result, many people turned to another kind of diet high in sugars (carbohydrates) that led to the epidemic of obesity and type 2 diabetes that has affected a growing number of countries since the 1990s. But *Dietary Goals* in fact called for a substantial reduction in sugar consumption and an increase in consumption of carbohydrates from fruits, vegetables and grains. In any event, in addition to sugars – not fruits, vegetables and grains – chicken was a winner.

Second, chicken is cheaper to produce than other animal meats. The broiler has been developed into a highly efficient meat-producing machine. According to some estimates, it takes 5 kilograms of grain to produce a kilogram of beef, ungulate land and water use requirements are higher than for chickens, and labor inputs are more extensive than for chickens. Chicken meat is produced at a 2-to-1 ratio of feed to bird. And the bigger, meatier birds that mature within six weeks are simply cheaper to produce than cattle who take months and years to reach slaughter weight. Finally, even if people in many countries are eating somewhat less beef, at times there has been an oversupply of chicken, which has helped to force prices down, and this in turn provides an additional incentive for those meat-eating consumers to buy more chicken product in all its forms.

The world's citizens are eating more meat. To put numbers on this meat, in the mid-1960s, according to the UN Food and Agricultural Organization, annual world meat consumption was 24.2 kg/capita by "carcass weight equivalent," in the mid-1990s this had reached 34.6 kg/capita, and in 2015 stood at 41.3 kg/capita, with a forecast of 45.3 kg/capita by 2030 – or nearly a 100 percent increase in 70 years. Of course, industrial countries are the major consumers, with consumption growing from 61.5 kg/capita in the 1960s to 95.7 kg/capita in 2015, or a 60 percent increase. Consumption has grown even more quickly in "developing" nations, even if it remains far behind wealthier countries: from

10.2 kg/capita in the mid-1960s to three times more in 2015 at 31.6 kg/capita. Much of the increase has come from poultry products, where, from the mid-1960s to the present day, poultry meat consumption has grown from 3.2 kg/capita to 17.2 kg/capita, over a five-fold increase, and with international trade very important in meeting demand. In this book, we focus on many of the leading poultry-meat consuming countries and areas that are – rounded roughly – in order: China in the first position at 19 million tons annually, followed by the US at 18 million, the EU at 14 million, Brazil at 9 million and Russia at 5 million.[4]

But in this book, I focus not on consumers, but on other important actors. They are the entrepreneurs who first recognized market possibilities for chicken meat, not only eggs; agricultural scientists who developed broilers; lobbyists who pushed for reasonable and favorable regulations; government officials at local, state and national levels, including legislators, and also personnel of various international agencies involved in research, standards and trade concerns; and animal rights activists, social reformers and moral critics and others who worry about the nature of factory farms.

Chicken analyzes the state of factory chicken farms in comparative perspective across the globe, including how chicken meat has become a major international trade commodity, with a focus on the major chicken nations. Readers will note some emphasis on the history of this industry in the United States. The reason for this is that the chicken CAFO in essence originated in the United States and has spread – like a farmed bird with wings – to the EU, Brazil and Asia, especially to China. No country has been immune to the pressure of industrial farming, and it is instructive to understand the nuances of its practices from one country to the next owing to greater or less sensitivity to environmental problems, questions of feeds and additives – including the use of antibiotics – how to deal with disease, efforts to keep costs down – perhaps at the expense of the welfare of animals, and farm laborers, and so on. In this comparison, one discovers that, almost universally, the greater the regulatory impetus to manage factory farms well, the safer, cleaner and more animal-friendly are the production facilities; the US is at the "less regulation" end of the spectrum.

It should also be remembered that there is ultimately little difference between one kind of animal factory farm and another: all are geared to generating meat as quickly as possible, minimizing inputs, uniformity in production from birth to slaughter, and result in similar environmental, social and other problems. Broilers are only one kind of chicken, and factory farming is only one way to raise animals. Urban farming has blossomed in a number of places. Backyard, humane raising techniques are proliferating, and chickens of a wide variety of breeds and purposes – meat and eggs – are raised in small-scale settings. But it is a relatively small number of chickens raised this way – hence, my focus on factory-farmed broilers. Here and there, I shall mention ungulates, pigs and other kinds of farm animals to highlight concerns about factory farms generally.

The chicken industrialization process is going on throughout the world, and this means that, if the United States may have been the originator of the chicken factory farm, then the other nations of the world – and the producers, regulators and consumers in those other nations – share in the moral, social and environmental problems created by the expansion of those farms. Unfettered capitalism is, in essence, the source of the factory farm: it is the driving force behind the industrial ethos of the broiler, and it is evident in the prevailing profit motive of the farms and in the logic of production. All of these countries therefore, to a greater or lesser degree, are responsible for the brutal, international system of food production that has resulted, and hardly the US alone.[5]

Chickens are treated as egg producers, meat producers, and dual-purpose types. The broiler – a meat producer – is most often a cross of the White Rock and Cornish breeds. There are others: red broilers, Delaware broilers (crossing Rhode Island Red hens with Barred Plymouth Rock roosters) and others. Plymouth Rock, New Hampshire, Langshans, Jersey Black Giant and Brahmas have also been introduced to the mix. And, finally, breeders have worked to make the broilers white-feathered. All this means that today's broiler is quite a hybrid animal, and very productive from the point of view of rapid muscle tissue gain. Other breeds do not reach meat slaughter age as quickly, so most operations go with the White Rock / Cornish breed. As will be noted below, the intensive breeding has led the broiler to be at risk for a variety of maladies, and particularly skeletal malformation and dysfunction, skin and eye lesions and congestive heart conditions.

The broiler made a long, scientific and industrial business trip over the century that is the focus of the book; early chapters consider the cultural history of the chicken and its "pre-industrial" history. An early bible of poultry published in 1914 indicated the growing importance of fowl to the US economy, well before production shifted to the southern states beginning from the 1930s. In 1910, Ohio, Missouri, Illinois, Iowa – mostly Midwestern states – were the major income producers from poultry, finishing with California in tenth place in income, while there were four New England states in the top ten in relative rank with reference to average farm income from poultry, with eggs the leading income producer. In terms of the number of poultry, the Midwestern states again dominated, with Iowa and its 23.5 million in first place.[6] Signs of centralized control of production that would characterize the industry from Brazil to China to India had already appeared, with transport innovations providing impetus – shipping in refrigerated and open train wagons to urban markets made this possible.[7] But, as yet, there was no indication of the rapidly coming consolidation, centralization and vertical integration of future years.

Stimulated by producers in the 1930s who saw cost-cutting possibilities in Fordist vertical integration, assisted by growing demand for chicken meat during World War II to bridge pork and beef meat shortages, and enabled by inattentive government regulation in the post-war years, the CAFO burst forth in the US in the 1960s and spread across the globe, beginning in the 1980s. In some countries, CAFOs are the major source of peoples' meat. Intensive animal production commenced in highly mechanized swine slaughterhouses, and in the chicken industry in several regions simultaneously, including Georgia and the south and Delmarva. Increasingly inexpensive feed (grain) and the growth of the transport industry also stimulated the industry.[8] Between 1950 and the twenty-first century, broiler production doubled on average every ten years. In 1959, US farms producing at least 100,000 broilers in a year accounted for 28.5 percent of production. That share doubled by 1969, and grew rapidly to the 1990s. Virtually all commercial growers now produce more than 100,000 broilers in a year, while the shift to larger operations continues – from 300,000 broilers in 1987 to 520,000 in 2002 and 600,000 by 2006.[9] To achieve such a dramatic shift in production and consumption, the US adopted the CAFO for cattle and swine, too, and in larger and larger factory farms that have, by the present, overwhelmed the countryside, local communities and the environment. Americans in 2015 consumed on average 80 lb (37 kg) of chicken annually, more than any other type of animal flesh. The US system of innovation, application and increases in productivity was followed everywhere, especially in China and Brazil.[10]

The US Environmental Protection Agency (EPA) defines a CAFO as an AFO (animal feed operation) that has been designated as a point source of pollution. The animals are confined and they are fed, rather than grazing on grass or other vegetation – at their own contentment and pace.[11] Yet the EPA had also made the determination that "facility" refers to a structure, and not to an entire farm. CAFOs are further defined by size. Large CAFOs have at least 700 dairy cattle; or 1,000 beef cattle; or 2,500 pigs if they weigh over 55 pounds or 10,000 if they do not; or 30,000 broilers if the AFO has a liquid manure handling system, or 125,000 if it does not. Medium-size CAFOs fall within intermediate size ranges and discharge wastewater or manure to surface waters, while small CAFOs are below the medium-size threshold, but are designated by local permitting authorities as significant contributors of pollutants.[12] For all livestock, the mean farm size has grown, and the "production locus" (number of head sold/removed) for over half of the broiler production in the US grew from 300,000 in 1987 to 520,000 in 2002.[13] At the same time, the EPA allows certain exceptions to the designation of CAFOs as a point source of pollution, enabling them to spread manure and other waste with inadequate controls, and that waste has polluted lakes, streams, rivers, ponds, wells and land far and wide. Nowhere in the world has the pollution problem been solved. This is ecological dishonesty, and, along with the replacement of small farms with large industrial facilities, it has destroyed communities and ecosystems. However you designate and define a big farm, there are huge quantities of animals and a great deal of shit, no matter whether the sheds are in North America, Asia or Europe.

Factory farms, like all successful new organizational forms in capitalism, attempt to maximize output from well-controlled – and minimal – inputs. In broiler production, as befitting vertical integration, firms called integrators own hatcheries, processing plants and feed mills. They contract with farmers to raise broiler chicks to market weight, and to produce replacement breeder hens for hatcheries. The integrator provides the farmer/grower with chicks, feed, and veterinary and transportation services, while the farmer provides labor, capital in the form of housing and equipment, and utilities.[14] In this way, the workers themselves are inputs. The chicks are inputs; the feed is an input; electricity and fossil fuels for ventilation, feeding, moving and heating are inputs; sheds, roads and machinery are inputs; and antibiotics are inputs. CAFOs also manage to push some of the costs onto the public that, sooner or later, are revealed to the public and require public suffering and expenditures to manage them.

One example of this phenomenon is antibiotics. The birds are at risk for a variety of maladies because of immune systems that cannot develop fully before slaughter. Industry turned to antibiotics both to prevent spread preemptively and to accelerate animal growth. Yet many of the costs involved in dealing with complex disease vectors on the scale of pandemics – for example, Avian Influenza – or to manage frequent outbreaks of *Salmonella* that require treatment of patients, often in hospitals, are borne by the public. Public health specialists worry about the growing antibiotic resistance of bacteria because of the overuse of drugs. Under greater and greater pressure from regulators and medical specialists, industrial chicken farmers have been forced to scale back the application of drugs somewhat. They and their spokespeople now refer to antimicrobials as a panacea for the problem. Recall that all antibiotics are antimicrobials, but not all antimicrobials are antibiotics. This is technically true, but also an Orwellian way to deflect the concerns of the public and regulators about the risks and benefits of antiomicrobials. If you need to use medicines in the production of meat, then is this not prima facie evidence that there is something wrong with the process?

A second area of concern examined in this book is the way that industrial chicken farming has become an environmental fiasco and public health outrage. Broilers are shit champions. They produce greenhouse gases from the methane in their bowels. For each kilogram (kg) of meat, roughly 500 grams of fecal matter result – no shit. Where is it stored? Whence the pollution and how is it spread? What of the offal? How hazardous and noxious is this material? What of the lagoons of shit and offal that result from the billions of animals (chickens and their meat-protein comrades – cattle, pigs and turkeys) throughout the world? Industry – and regulators – have been slow in response, and the dangerous, bubbling liquid masses – or the dried, odiferous "cake" that is treated by industry as manageable – have spread across the landscape.

Chicken CAFOs, beef CAFOs, pork CAFOs and other such factory farm operations are dreadful ways to mass-produce animal meat as if it was like any other commodity that can be mass-produced. They are a worrisome example of how the capitalist impulse to profit while meeting consumer demand has a very dark side: animal cruelty, worker exploitation, pollution and so on. Similar systems exist for other kinds of animals and animal products that indicate the universal nature of the meat commodity machine. One example is the tiger and bear farms of East Asia that enable rife animal brutality, where many consumers do not care about that suffering, and where powerful states that could regulate or prohibit the industry do nothing. They tolerate abusive practices, and even promote or ignore them in the name of money-making.

The persistent and long-lived trade in bear gall bladders and bear bile, for example, threatens the Asian bear species.[15] While this trade is legal within some countries, cross-border trade of bear bile products is prohibited by the Convention on the International Trade in Endangered Species of Wild Fauna and Flora (CITES). But it continues and has changed from being purely for traditional medicinal to providing a commodity, with bile now being found in such products as cough drops, shampoo and soft drinks. A great number of countries buy and sell bear bile products originating in other countries in violation of CITES: Myanmar, Hong Kong, Laos, the Republic of Korea – the latter often with products from wild bears in Russia where hunting and trade of them are legal.[16] The bears (and other animals in this trade for parts) are kept in miserable, caged, claustrophobic conditions – roughly 20,000 bears alone, across East Asia.

There has been progress in raising consumer and producer awareness of the cruelty and immorality of farms in some places – for example, Vietnam and Korea, which have promised to close them by 2020. Yet China remains wedded to them and is unwilling to entertain closing them at the highest levels of government, among consumers and, of course, the producers.[17] Bear farming in Laos has begun to shrink, but the growth of facilities in the northern part of the country under private, mostly Chinese, ownership counters that trend.[18] A European Parliament resolution of 2006 calling to end bear bile farming in China fell on deaf ears as China rejected this interference in a domestic issue,[19] while tiger farming in Laos also supports primarily Chinese interests, tastes and consumers in the sale of parts, teeth, claws, paws, and meat.[20] But our focus is the broiler factory farm. Suffice it to say that broiler meat, too, is traded internationally, with birds kept in miserable conditions, although not in violation of CITES – because, with billions of the fowl, they are hardly an endangered species.

Conveyor-Belt Chickens

This book draws on a number of fields and approaches to write a history of the rise of the broiler in an international context in the twentieth century, although it is largely political history, environmental history and history of technology. In their explorations of the relationship between technology, humans and nature, several scholars and journalists have focused on the economic, political and technological factors that have a significant role in the transformation of farming – generally, and in specific animal husbandry sectors – into an industrial project. They write about animals and domestication, farming and industrialization, animals and research, animals and globalization, and so on, each with a unique and important perspective in such genres as women's history, labor history, business history, history of science, anthropology, history of technology and environmental history. They ask: what role do natural objects play in society and when do natural objects become technologies?

William Boyd argues that the "subordination of the meat broiler to the dictates of industrial production" indicates how technological change in agriculture further blurs the distinction between nature and technology.[21] Focusing on broilers, he considers how they were incorporated in the technology and political-economic system. Boyd writes, "tethered to innovations in environmental control, genetics, nutrition, and disease management, the industrial broiler emerged as a vehicle for transforming feed grains into higher-value meat products." Like other such products, the broiler not only transformed food production – and diet – but "facilitated a profound restructuring of the relationship between nature and technology."[22]

Deborah Fitzgerald has demonstrated how biological organisms have been remade into agricultural commodities, with the production of scientific knowledge and the transformation of that knowledge into commercial practice. In this process, practice has become increasingly industrial, large-scale, profit-oriented and intensive in production. In *Every Farm a Factory*, she describes how businessmen, government officials, rural lenders, farm management specialists, engineers and extension agents imparted an "industrial logic or ideal" to agriculture after World War I to tie farmers into an increasingly integrated national system of production and consumption. They were pushed by market forces and by the industrial logic of rationalization and standardization. If farmers did not embrace the ideal of industrial logic, then their use of industrial methods made them part of the system. They bought into tractors, then worked with bankers who encouraged them to buy more machines, and then found themselves pushing the land to pay for the machines, and turned to specialization to produce cash crops. Factory farming continued and has expanded to this day, and broilers enable us to follow its continuing transformation.[23]

In another work, Fitzgerald argues that factory farms received impetus from science at land grant universities, and from companies that sold science – in the form of seeds – to the farmers.[24] Ultimately, it appears that government-sponsored agricultural research and its dissemination from extension services, both of which were paid for by taxpayers, helped not so much individual farmers, but large companies that came to dominate agriculture in a variety of fields – soy, corn and now animals.

Several studies – and there are many, many more than I mention here, including outstanding investigations of CAFOs – pointed the way for this book, and to my understanding of the chicken. In the readable and informative *The Chicken Book* (1975), Page Smith and Charles Daniel offered a biological, zoological and cultural history of the domestic chicken from domestication. They criticized chicken factory farming – in particular, the battery-cage system of egg production.[25] In a book about several different meat industries in historical perspective, Roger Horowitz discusses how manufacturers in the twentieth century managed to standardize animals from the field to the consumer in the mechanization of meat production; he includes a superb chapter on the chicken. Horowitz urges us not to succumb to the belief that the victory over nature has been complete, but to recognize a series of problems of race, gender, safety and public health that persist to this day.[26] In *Big Chicken*, Maryn McKenna discusses how the modern chicken industry is both founded on antibiotics to accelerate weight gain and reduce losses from infectious diseases, and needs them to deal with the conditions it created, which enabled the spread of such foodborne illnesses as *Salmonella*, and superbugs such as *E. coli* with the MCR-1 gene, that are difficult, if not impossible, to treat.[27]

Ruth Harrison's *Animal Machines* (1964) criticized nascent factory farming, revealing the suffering of animals at the hands of handlers and their machines – for example, calves in veal crates and birds in battery cages. Harrison helped shape public opinion about factory farming and the need for animal welfare, triggering a series of legal reforms. Harrison's book is no less important in the twenty-first century, since these farms have spread all over the globe. In some ways, *Animal Machines* is to animal welfare what Rachel Carson's *Silent Spring* (1962), written at roughly the same time, has become for the environment, and continues to be important in urging humans to consider the lives of now billions of factory-farmed meat animals.[28] Karen Davis, of United Poultry Concerns, has also documented in great and disturbing detail the need for "compassionate and respectful treatment" of chickens in a series of book and other publications.[29]

Annie Potts focuses precisely on the chicken in a cultural and social history of the bird, and includes a chapter on chickens as meat machines in a readable and informative study.[30] In a handsomely illustrated natural history of the chicken, Joseph Barber provides chapter and verse on the chicken historically, but mostly from a sociobiological and behaviorist perspective.[31] These books reflect the growing concern about how chickens became meat machines, and how – tracing the cultural history of the bird – we might recover some of our humanity in recognizing its place in our global world over the centuries and millennia.

My book also engages environmental history. How can it be otherwise with such a topic? In a series of important studies, the Pew Charitable Trusts researchers note how factory farms have changed the discourse on environmental risk, changing what had largely been sustainable agricultural practices in significant ways, even taking into account growth in population and consumption. The new farms focus "on growing animals as units of protein production." They import feed, they add medicine, all to get animals to market weight as quickly as possible. They overlook the "natural productivity of the land."[32] The question is how to sustain fertility of the soil through conservation, not driving it to the ground, and to ensure local food security; how to produce healthy, non-toxic food; how to ensure good salaries with social support in rural regions; and how to respect the goodness of animals and the environment. A variety of other foundations and organizations have highlighted growing concerns with CAFOs.[33]

In the opus of her work, Harriet Ritvo reminds us of the importance of the subject of animals in environmental history. She points out that environmental historians closely examine the history of livestock and domesticated animals, not only for the impact of human–animal relations on the environment, but because animals are connected with various institutions, including research institutes, agribusinesses and multinational corporations that seek to make and patent them and their feeds. They include hunters, trappers and furriers with the rise of commercial interests and overexploitation; not only farmers, but breeders, scientists, and researchers. They are local people and consumers at supermarkets. They are connected to granges and cooperatives and extension services."[34]

Several scholars note how the production of livestock is no different from the production of many other products, within and outside of the agricultural world – for example, automobiles – with specially designed buildings to maximize controlled space, minimize input and significantly increase output. For livestock, special buildings, barns and outbuildings, motors and conveyors have entered agribusiness. The products include chickens, pigs, cattle – both dairy and beef – and so on. As Susan McMurry notes, their industrial production both benefits from advances in public health and becomes entrapped by them. For example, advances in bacteriology changed dairy production, with government sanitation officials pushing regulations to ensure safe milk production as it was transported from the countryside to the city, and with rising milk consumption as a substitute for human breast milk. Perhaps pasteurization, bottling and cooling to standards was the only possible outcome. She notes that barns, trucks, highways, local plants "entered the mix of places where bacteria might grow in milk." For the barn, McMurry writes, "metal ventilators sprouted from the roof ridge; milk houses appeared; hog houses were demolished or moved; poultry houses were relocated; new privies were built; water systems were installed (at least at the barn and the milk house); and new horse stables were built." Eventually, human handling was reduced to a minimum,[35] as it has been in the case of broilers. Indeed, chicken meat production reflects all of these tendencies. Even the architectures of chicken production reflect the considerations of maximum efficient use of space and reliance on modern inputs of food and drugs to make those spaces work optimally. In many places below, the reader will have the opportunity to consider the way in which technological advances seem to impel the factory farm onward – what I refer to as a technological imperative that suggests a determinist argument.

Broiler production of similar chicken units is essentially no different from the production of monocultures of various other plants – bananas, coffee, rubber and so on. These living things are based on the drive for manageable units of output based on industrial understandings, and the belief among promoters that they can prevail over climate, seasons, terrain – whatever the physical, geophysical or biological problem. As Richard Tucker demonstrated, the American economic strength in Central and South America, the Caribbean and the Pacific Rim had significant impacts on local environments and people through colonial, plantation and post-colonial production of sugar, fruit, coffee, rubber, cattle and timber. American business interests, with the help of government, sought to establish monocultures of bananas, rubber and other commodities, using slaves or indentured labor, armed with dangerous chemical pesticides, herbicides, fertilizers that eventually polluted water and soil, and also antibiotics. He notes how financial interests pushed the creation of these monocultures.[36] Investment went into production and extraction, not the development of civilian infrastructure – roads, hospitals, stores and schools. Often with the assistance of local and national officials, they have pushed the monocultures with the promise of local benefits, and yet the local people suffer the burdens of production, social disruption and environmental change.[37] In the modern chicken industry, similarly, and around the globe, local producers – contract laborers – and their families work in difficult conditions for low wages, while big businesses far away reap the harvests and push the costs of environmental and social disruption onto those laborers and their communities.

Some people have written about the industrialization of agriculture as natural and expected, if from a technologically determinist and nearly utopian perspective, ignoring the costs and consequences, and suggesting that local communities will always adjust. Hiram Drache, a historian of agriculture, writing in the 1970s, insisted that larger-acreage farms were the most efficient and modern of American farms, while noting that family farms, the mythical foundation of American republicanism, would survive the onslaught of technological change. By efficient, he meant by such measures as acres harvested per machine, yield per acre, and yield per animal. He did point out an important fact: far from being a product of capitalism alone, government programs were central in stimulating the growth of large-scale agriculture,[38] as they had been directly and indirectly through the US Department of Agriculture (USDA). Powerful machines – greater horsepower – enabled one farmer to do the work with fewer hours of hired labor per season, while comfort at the controls – two-way radio, air conditioning, a smooth ride – enabled expansion of farms to the horizon. Drache argumentatively suggested that government programs of "a non-price support nature, such as Occupational Safety and Health standards, environmental regulations, and social and labor legislation" were inappropriate, for they would discourage the small-farm operator from staying in the business as too expensive.[39]

Yet Drache found it possible to conclude that, even if large-scale practices were advantageous for all meat-animal industries, the farms of the twenty-first century would still be family-oriented units.[40] On this level alone, Drache ignored the fact that massive farms armed with industrial tools, lax regulation, and government subsidies do not constitute "family farms." He optimistically noted that the social implications of the tying of industry to agriculture would be substantial, but insisted that "people will adjust and the end result will be a better life style." He tried to suggest that people who protest against this situation are Luddites of some sort, like those who railed against the Industrial Revolution where all turned out for the better.[41]

Yet the family farm has for a long time not been a dominant political and economic feature of US agriculture, nor of that in many European countries, although the myth of these family farms persists, and, wherever CAFOs appear, the smaller producers seem to suffer.[42] Indeed, family farms are physically smaller, have lower average income, an increasingly small share of overall production, and receive fewer benefits and subsidies than larger farms, including those privately owned or run by absentee owners and corporations. In fact, the top ten "farmers" in the US in 2018 receiving subsidies were corporations with an average annual subsidy of $18 million each.[43] In Europe, the same "myth" of family farms as being somehow, by the twenty-first century, stable economic and social units of production prevails.[44] Rather, the factory farm dominates in such forms as the CAFO, and they are a shitstorm of "inevitable progress." We cannot ignore the human, social, biological and environmental costs of the factory farm any more than the pollution, horrific social trauma and maimed and killed workers of the Industrial Revolution.

This book cannot give full attention to the social history of chicken factory farms, both because the subject requires its own complete study, and because the chicken itself is our focus. But the chicken itself extends far beyond the fields and broiler sheds to the homes and farms nearby, to the local banks and government, to social services and infrastructure, all of which seem to collapse under the weight of CAFOs – and smell none too good either. In CAFO farming, tautologically and dangerously, large farms dominate, and where there are large numbers of farms, the larger ones by far produce more animals. Their only concern is output of chicken units. CAFOs respond to shareholders and CEOs and other investors who are distant from the surrounding towns and the people in them. Who cares for and tends to the animals? These people are contract laborers, or migrant workers, who rarely receive such sufficient social benefits as insurance, and who face great financial uncertainty and challenging physical labor. It may be that Europe's safety net makes a big difference for CAFO workers, but in most of the world these farmers live on the edge of economic uncertainty.

This is a hard life, dominated by obligations to integrators that control virtually all of the inputs, and the margins for success or failure are slim for the labor contractors, especially considering capital costs. A pair of broiler houses, with automated equipment for feeding and watering the birds, and climate control systems, mechanized equipment to gather broilers for shipment to processing plants ("chicken harvesters") and to remove litter from the houses, can cost from $350,000 to $750,000. Broiler houses built in the last decade cover 20,000 square feet (40 feet wide and 500 feet long; approaching 1,900 square meters). In an average year, a single house might produce 115,000–135,000 broilers; few houses built recently are less than 20,000 square feet. Some grow-out operations have up to 18 houses, and this enables continuous production when some sheds undergo litter removal and upkeep.[45] This is a radical change from the much smaller operations in the 1950s and 1960s, and who can afford these costs and these sheds?

The writer Paul Crenshaw gave a sense of the social costs of chicken factory farms when writing about the industrial transformation of Arkansas into a chicken coop. Arkansas, the home of Walmart, the largest company in the world by revenue, that sells inexpensive Chinese and other goods under the banner of "Made in America," is also the home of Tyson, the largest chicken operator in the world. Crenshaw observes that Arkansas roads and highways are bordered with chicken factories and packing plants, and filled with trucks carrying birds – live for growing, and dead for sale. Beyond the strips of natural beauty, near streams and creeks and rivers, workers in factories push and prod the birds to maturity, and cut and drain them, transforming them in seconds into food. The guts and shit truly have no final resting place, but fill the air with acrid odor and the ground and water with toxic mess. Crenshaw refers to "gut trucks" that "weave along roads, leaving a swath of olfactory offense in their wake." He notes the "chemistry lesson" required to understand the grotesqueries of decomposition.[46] Crenshaw chronicles the scale of chicken houses, 100 yards long with 25,000 bird residents, tended to by poorly paid laborers who are cleaning, checking, carefully controlling lighting, fixing ventilation, regulating temperature, navigating rodents, maggots, flies and "the dead pits." The dead pits, covered with concrete slabs, are cauldrons of crap. These houses are teaming with motion, all of it natural – yet none of it natural.[47] No longer do chickens hunt and peck, find worms and bugs, and lounge in the shade. They are pushed and prodded, vaccinated and fed, in an artificial environment that limits their aggressiveness, packed tightly, to grow in vertically integrated factories like those of Tyson, in sheds like those of Tyson, to train their movements entirely to fattening and death. When they have been evacuated from the houses, laborers enter to fight the accumulated smell, feces and urine, fumigate and prepare the sheds for more sweet little chicks to begin the transformation into meat machines.[48] The chickens shit in their food, and Crenshaw suggests that the way we raise them means that we defecate in our own food, too.

From Cage to Carcass

This book aims at a fuller understanding of the chicken raised in factory farms, perhaps as a Neo-Luddite might write it, hoping to promote realization about how industrial forces and capitalism changed what was a domesticated yard bird into a protein machine. If this can lead to greater regulation of the industry in the name of animal welfare, pollution control, public health and safety, then I will have succeeded in some small way in meeting my goal. But, above all else, this is a political and environmental history of the broiler, from its early domestication as a chicken, to a provider of the occasional egg or a one-time tough and sinewy meal, or as a meaningful religious symbol; to its appearance as a friend of the family, the farmer and as a collector's item – Queen Victoria of England being among their admirers; to the factory farm. And it is a history of the chicken that has been transformed by humans from an active, social bird with an ecology of running, pecking and establishing a social (pecking) order that required sun, air, greens and exercise, to one entirely confined to a vanishingly small space, and intended to be chowed down after assembly-line execution before even reaching full maturity.

Through hybridization, and eventually through genetics, capitalism was joined with research and agriculture to build broiler birds with specific growth, fat, meat, enzyme, flavor and other properties. By 2000, just three firms in the world provided the vast majority of these magnificent meat machines through their breeder banks. According to the United Nations Food and Agriculture Organization (FAO), there has been a marked decline in the past half-century of farm livestock breeds, in large part because of this process. "Up to 30% of global mammalian and avian livestock breeds (i.e., 1,200 to 1,500 breeds) are currently at risk of being lost and cannot be replaced."[49] A Purdue University study in the *Proceedings of the National Academy of Sciences* reported that 50 percent or more of ancestral chicken breeds have been lost, and that the greatest decline in chicken diversity took place in the 1950s with the introduction of industrial chicken production.[50]

In the late 1940s, encouraged by the Great Atlantic and Pacific Tea Company (A&P Supermarkets), thousands of farmers took part in "Chicken of Tomorrow" to present the best, meatiest breeds, which A&P intended to use in display cases around the nation to build America's taste for chicken. These breeds have given way to the technologically superior broiler (meat generator) of factory farms: a faster-growing chicken with more meat on its bones. The birds are strong in meat, but, like any monoculture, prone to disease, foodborne and otherwise, and with skeletal, vision and other weaknesses. In many ways, they are an icon of CAFOs: meatier, but sicker, birds requiring more and more energy inputs, including antibiotics. They are chickens with an industrial essence. It is time for a new contest – a contest not for a new Chicken of Tomorrow, but rather for a new kind of agriculture, one that is less focused on corporate profits and more focused on producing strong healthy farms and food, that strengthens farm communities and supports local as well as distant markets.

Panopticon of Production

This book will follow the chicken historically from its farmyard frenzy into its bondage as a broiler. Chapter 1 celebrates the multicultural manifestations of the chicken as a glorious bird, its commonalities with humans (love of life, happiness in brooding and its manifestation as one of god's creatures in a variety of ways) and its celebration in literature, art and music, from religious sacrifice and cockfighting to a plaything of the wealthy, and always as a hunting and pecking farmyard friend, even if destined for consumption, from domestication to the eve of the twentieth century.

Chapter 2 explores the ecology of chickens and how ideas and understandings of chicken behavior, health and habitat have significantly changed – perhaps it is more accurate to say "have been changed" – since the turn of the last century. Like others who have studied modern-day agriculture, I suggest that an agro-industrial imperative pushed these concepts from a natural to a technological foundation. In the setting of the factory farm, the question is whether the broiler is even a bird any longer. The manufacturers want to have it sautéed both ways – as a chicken to feed us, but also as an industrial object to be regulated like an automobile, put together from various parts, and yet without what they see as onerous regulations, because this is only a bird.

The rapid industrial transformation of the chicken from an animal well adapted to its natural worlds – from Southeast Asia to the Savannah of Africa, from the backyard farms of the plains states to the peasant farms of nineteenth-century Europe and Asia – to an industrial object to serve entirely as meat or egg layer is the focus of chapter 3, where, drawing on an eighteenth-century French doctor and *philosophe*, I call the broiler a "machine" and identify its many parts, and the increasing use of a variety of genetic, chemical, electrical and other inputs that, being employed more and more from the second half of the twentieth century, completed the transformation of the chicken from a barnyard animal to a factory farmyard animal. In this chapter, we also examine the growing use of antibiotics and other such chemicals, the role of regulation and inspection in avian-industrial safety, and the growing scourge of foodborne illnesses and pandemics.

Chapter 4 explores thematically the kinds of environmental problems that arise with factory farming – in this case, with broiler production. There's a lot of shit to be tabulated and estimated and weighed, and a lot of other pollution as well: methane, run-off, heavy metals, antibiotics, land use and the like. Rather than provide that tabulation, however, I offer discussion of how factory farms have evolved to be such environmentally unsound facilities, the weakness of the regulatory impetus, and the externalized costs that consumers pay and will pay for remediation. Greater and greater awareness of these costs and problems dates to the middle of the twentieth century.

Moralists, economists, citizen groups, regulators and many others are pushing back against the human, animal and environmental costs of factory farms. Some of them recognize the public health costs are already so great that the world's meat eaters must act today in order to have their chicken and eat it, too. In chapter 5, we explore historically the rising protest against these farms. I define protest broadly to include government intervention in the production processes through inspection and regulation dating from the late nineteenth century. Protest involves a variety of NGOs and other organizations concerned with environmental questions and animal welfare. One might argue that, in order to continue as meat eaters, we must first of all consider animal welfare, lest we cheapen ourselves by permitting wholly industrial processes to overwhelm our morals, sensibilities and taste buds. We should hear the sounds of nature – and make our own sounds of protest – even while smacking our lips.

If chickens left the forests of Southeast Asia millennia ago to enter the human world of domestication, then they have taken flight again as products of multinational corporations. Chickens – dressed, processed, cut-up, wrapped, frozen and so on – are an international commodity. One in five bits (or pieces, or kilograms) of chicken is exported somewhere, often thousands of kilometers away. Global sales of fresh chicken shipments reached $6.6 billion in 2018, and frozen-chicken international sales were $16.1 billion. In chapter 6, we examine international trade and several of the major countries' industries and producers. Trade is based as always on domestic production and policies, and on international rivalries and demands. Russia and the US are having a fowl war. They are not alone in battles over tariffs and imports. And international trade also involves great risks of introducing or reintroducing highly pathogenic Avian Influenza A (H5N1) to uninfected countries. Whence are your chicken nuggets and wings?

A brief epilogue intends to make sense of this analysis of the panopticon of production, and offers a section on many of the chicken metaphors, sayings and puns I have been able to find – or, in the case of the puns, to generate – in order to avoid sprinkling them too liberally on other parts of the book, with only pieces and nuggets of them coming beforehand. Advocates of factory farming will argue that broiler food is inexpensive, wholesome and environmentally sound. This book may demonstrate otherwise. And, on a final note: I am an omnivore. I like my chicken. But this is food not to be eaten lightly. The chicken is a domesticated animal, a cultural and religious artifact, now turned into a Frankenstein fowl whose future life is unsustainable. The following chapters explore chicken–human interactions, from the first domestications and cultural manifestations that have persisted to this day in art and music – including in cock fights and beauty contests – to regularized farmyard activities, and finally to systematic exploitation of the broilers' extensive benefits to humans as a source of protein, which have been increasingly facilitated through industrial production and international trade. The chicken that originated as a domesticated bird in Southeast Asia has returned, chilled and chopped-up, as a commodity for those very Chinese, Japanese and other farmers and consumers.

2 *This nineteenth-century chicken poultry stock illustration indicates the diversity and beauty of various birds.*

1

Chicken Culture

> A Jewish woman had two chickens. One got sick, so the woman made chicken soup out of the other one to help the sick one get well.
>
> – Henny Youngman

Chickens have millennia-long proximity to humans. Chickens, like dogs and cats, have lived with us in our houses, huts and barns, outside in the yards, and nearby in the fields. They become friends and playthings, not only meat- and egg-providers. They have been the focus of music and poetry, the objects of art and sculpture, and the center of attention of clubs, all of which are discussed in this chapter. Chickens, as well, reflected norms and expectations of race and gender in the way they were raised and who raised them. People keep chickens until they become too old or lose their egg-producing capabilities, or until the moment they have decided that it is time to eat them. Often, they grow fond of their animals. My Maine "down east" cousins raised a pig, whom they named "Pig," they told me, to lessen the sadness of parting with her at full growth for her to become sausage, pork chops and so on; they stopped raising pigs.

The world's inhabitants produce the prevailing share of today's meat animals in industrial settings, and in so doing deny them individuality in behavior, and even, at times, any sense that they are animals. People have no intention to stop raising them. Factory farms remove them from the consumer's awareness by rearing them in boxes, sheds or corrals, big and small, before hurrying them to conveyor-belt abattoirs. We can see the change in slaughtering practices on canvas. Bernando Strozzi's *The Cook* (1625) features several fowl in the cook's hands, hanging upside down by their legs, lying on a nearby table, ready to be plucked, eviscerated and baked. In Pehr Hilleström's *En Qvinna slagtar höns* (A Woman Slaughtering Hens), a woman, with help, is holding a chicken between her legs as she cuts its neck; several others, already dead, lie at her feet (c.1775).

Since domesticated, millennia ago, chickens have been cherished, celebrated and collected for their familiarity and even commonness, their suggestion of fertility, and their beautiful feathers and carriage by wealthy folk, including Queen Victoria; used in cockfights by wealthy and non-wealthy people alike; and often met their end in religious sacrifices at the godly hands of Jews, Christians and Muslims. Marc Chagall's *The Cock* (1928) reveals the love between the animal and a harlequin riding her, while Pablo Picasso's *Rooster* (1938) looks like an American weather vane. Painted Ukrainian Easter eggs communicate traditional meanings and sacral ones, including fertility; my favorite painted Easter eggs are from the Zaporozh'e region. Similarly, musicians and composers have engaged chickens. Rock musician Alice Cooper once received a live chicken on stage from his audience and threw it back to the crowd, where, apparently, it fared badly. Sergei Prokofiev briefly owned a small electric incubator that gave him pleasure when the chickens were hatched, but their overwhelming sickliness led him to give the incubator away. The magical glowing firebird of Russian folklore and of Sergei Stravinsky's *Firebird* ballet is, however, not a chicken.

I have found no references to chickens in Shakespeare, although it is worth remembering that "'tis an ill cook that cannot lick his own fingers."[1] Elsewhere in the medieval European world, chickens were the source of metaphors and sayings, although not entirely in keeping with their bright, curious and seemingly optimistic outlook. The expression, "A bad chicken was brooding," was noted in fifteenth-century Flanders when the guilds of Ypres were in open revolt to regain the privileges lost to the ruling patrician class.[2] In the *Canterbury Tales*, Geoffrey

Chaucer used food as staples of everyday life and the eating habits of various characters to convey their attitudes, values, level of wealth, and the variety and abundance of food in the fourteenth century. In several poems, animals play a significant role, but in only one, as far as I can tell, does a chicken or rooster provide us with a lesson. In the "Nun's Priest's Tale," the rooster, Chanticleer, manages to escape the jaws of a fox after first recklessly allowing himself to be caught by crowing proudly with his eyes closed.[3] So don't be reckless, watch out for foxes, and stay out of factory farms.

The chicken has long been holiday meat at the dinner table. According to the chef-authors of one cookbook:

> Without chicken, humankind would never have become the species it now is: the top of the food chain. And vice versa. Alongside humankind, chicken has spread all over the world and its diversity has increased enormously. For thousands of years, the range of human cultures has not only found its counterpart in the biology of the chicken, but certainly also in the nutritional importance of our most important 'companion animal.'[4]

US President Herbert Hoover – famously and mistakenly – forecast a time of prosperity in the US during the 1928 presidential campaign, in a Republican Party campaign brochure that claimed that, if he won, then there would be "a chicken for every pot and a car in every garage." He did win, but the stock market crash of 1929 "plunged the country into the Great Depression and people eventually lost confidence in Hoover,"[5] and showed Republican claims of unregulated capitalism as the savior of the common citizen to be an empty pot. It would be another two generations before Americans were consuming as much chicken as Hoover anticipated. Republicans in the US Congress in the twenty-first century have spent significant energy trying to cut school lunch programs in America to ensure that no one gets chicken.

Chickens serve more than literary, cultural and religious purposes. For centuries, people have collected chickens for display because of their beautiful feathers, crowns and bodies. They have raised them as pets. They have used the roosters for cockfights, equipping the hapless birds with spurs to ensure bloodletting and death. The latter has a 6,000-year history. In his *The Commendation of Cockes, and Cock-fighting; Wherein is shewed, that Cocke-fighting was before the coming of Christ* (1607), George Wilson tried to justify the practice by showing its ancient roots. He wrote, of Henry VIII:

> He did take such pleasure and wonderful delight in the cocks of the game that he caused a most sumptuous and stately cock pit to be erected in Westminster, wherein His Majesty might disport himself with cock fighting among his most noble and loving subjects who in like manner did affect that pastime so well, and conceived so good an opinion of it . . . that they caused cock-pits to be made in many cities, boroughs and towns throughout the whole realm.[6]

Cockpits spread in the seventeenth century among such self-proclaimed civilized people as the Brits who decried the savages they had conquered in their empire, yet who provided calendars of forthcoming bloodlettings, and who organized free-for-alls among several prize birds.

If cockfights have been outlawed in most places of the world, they still continue, and they draw blood-hungry crowds and gamblers to see one cock's spurs dice another cock's neck. In India, they were transformed from a pastime of elites into that of the common folk after British colonization and the adoption of more "gentlemanly" pastimes such as cricket and tennis.[7] Anthropologist Clifford Geertz encountered cockfights, which were generally illegal in Indonesia, when he was doing his doctoral research. He claimed, based on interviews with local people and observations of the cockfights and the men running them and betting on the fights, that the cock indeed, in Balinese as in English, has the double-entendre meaning of a prideful, powerful man, and the penis, a phallic symbol, and also that the rituals of the betting and fights reflect important social relationships of kin and village governance. Cockfighting was, in fact, not only a traditional practice but a religious ritual. The spilling of blood on the ground warded off the evil spirits. Yes, some cockfighting occurs as a business for entertainment. But its traditional nature has encouraged local police often to look the other way.[8]

In the US, the state of Louisiana outlawed the vile practice only in 2007. In 2000, the then Senate majority leader, Trent Lott

(R–Mississippi), a notorious segregationist and a fan of bloodletting, blocked a federal law that would have ended cockfighting in Oklahoma, Louisiana and New Mexico – perhaps because he had no moral qualms about allowing violence toward other creatures.[9] This book asks, implicitly, whether the slaughter of chickens in CAFOs is not as immoral and unwelcome as cockfighting. Surely, the unwillingness to regulate CAFOs more carefully, from animal rights, public health and environmental points of view, reflects a kind of thinking that birds have fewer or no rights compared to the consumer. Also, likely the greater reluctance to regulate CAFOs (compared to cockfighting) reflects the greater economic power of those who profit from them.

Chickens also mean a lot in other kinds of sport. Russian and American officials are having a cockfight over American exports to Russian markets, while Russian officialdom had no love for chickens inside World Cup soccer stadiums during the 2018 championship. Authorities in Kaliningrad, Russia, forbade fans from bringing live chickens to matches. Some football fans dye chickens in the national colors as a good-luck symbol, including those from Nigeria, who played Croatia in Kaliningrad. The regional culture and tourism minister, Andrei Yermak, said that "fans from Nigeria asked whether they could bring a chicken to the stadium. It's their symbol and people support the team with them at all the games. [But] We told them they cannot bring a live chicken at all." Yermak advised Nigerian fans to call a hotline to learn "where to buy a chicken. We're prepared to satisfy even the most eccentric requests."[10] But what is eccentric about ubiquitous chickens?

Chickens and World Culture

I cannot pretend that the special bond between people and chickens is more significant than that for other forms of livestock, let alone pets. Yet a brief review of this interaction reveals their importance as friends, for economies of production and forms of farm labor organization, and also the fact that chickens generally were easier to care for than most quadrupeds and made fewer demands for food and shelter on their owners.

Chickens have long been significant for people on the edge – peasants engaged in subsistence agriculture, female laborers on the farm, and such people as Russian Jews who were forced into difficult emigration by the anti-Semitic Tsarist ideology of "Autocracy, Orthodoxy and Nationality." And the tsars indirectly contributed to chicken farming in the US. Alexander II and his secret police, the authors of *The Protocols of the Elders of Zion*, the sickly fake news of anti-Semitism, terrorized Jews who lived in the Pale of Settlement through state-sponsored pogroms in the 1880s, setting off a wave of immigration to the United States. By 1915, more than 3 million people had arrived, many to New York City's Lower East Side. Many Jews also settled in agricultural communities and became the nation's egg providers. A number of Jewish philanthropies outside Russia supported emigration – for example, the Jewish Colonization Fund and the Alliance Israélite Universelle in Paris, with the former providing training programs and land outside the Pale. The Rothschild family paid for a series of irrigation projects in Palestine in the mid-1800s to support nursery, vineyard and flour mill activity; the Rothschilds also bought land for colonization. Most of the settlers went to the cities, but an important group became part of an experiment in Jewish agricultural colonies, including Southern New Jersey's Alliance and Woodbine settlements. Both were established to work the land and both contributed to egg production.[11] Many of the immigrant agriculturists settled farther away, in colonies from the Dakotas in the North to Louisiana in the South, and to California in the West, often without any well-conceived plan or forethought, apart from setting up poultry and dairy farms.[12]

They contributed to the rise of the egg industry in California, north of San Francisco in Sonoma County and in the Sacramento Valley. Industrial, mining and agricultural development went hand in hand. The poultry industry began to develop along with the California gold rush in the 1850s and 1860s, and was complemented by Jewish families who had escaped violent, official discrimination in the Russian Empire. By the late 1880s, many people raised poultry exclusively near Petaluma, if mostly for eggs. There was sufficient shade and plentiful water in most places. By 1905, 1 million laying hens squawked in the surroundings – the farther out from the city center, the bigger the farms, some reaching 500 to 600 acres at 10 to 15 miles out. Fowls were shipped alive in coops with wooden frames and wire rods or heavy hexagonal-mesh wire netting, while eggs were moved in specially built heavy cases holding

36 dozen eggs, the majority of which were marketed in San Francisco, although some went to mining districts. Government orders were crucial to the business, with shipment on steamers to Alaska, Hawaii and the Philippines. Most of the hatching and raising occurred using artificial methods; a large incubator plant always had some 2,500 eggs in the course of incubation.[13] Mechanization was already a hallmark of production at the beginning of the twentieth century:

> There is here a chicken hatchery which is believed to be the largest in the world. It consists of an incubator house in which 30,000 eggs are in all stages of incubation; two houses, each 300 feet in length, and each having a capacity of 2,500 laying hens; and two brooder houses, each 160 feet in length, and having a combined capacity of 100,000 broilers a year. In this plant a ton of feed is used at each feeding time. Electric cars are used in the buildings for carrying feed and wash water and for collecting the eggs and the offal. The daily gathering of eggs is about 3,600. The feed is all stored in the upper floor and is delivered into the feed troughs by means of chutes. Water is furnished by a system of pipes to each pan. The floors are all of concrete, and the whole institution may be whitewashed in three hours by the use of machinery.[14]

In many societies, peasant women were the poultry farmers. Back in the late-nineteenth-century Tsarist Empire, in an agricultural sector of the economy buffeted by drought, hurt by inefficient land-holding practices, and suffering from backwardness in terms of techniques, ideas and equipment, poultry keeping was crucial in providing a margin of safety to families, and additional on-farm income to poverty-stricken peasants. It was the preserve of female peasants, who recognized the benefits of improved poultry breeds. The spread of the railway enabled poultry products to begin to move beyond the village to more distant markets, including for export.[15]

In Ireland, too, women were largely the handlers of poultry. By the late nineteenth century, rearing poultry for sale rather than household consumption had become an important occupation for farm women. The receipts for eggs often helped to pay rent and buy groceries. One Irish MP said that women "were more successful poultry-rearers than men because they understood the temperamental character of the fowl and so were 'more careful about the little details.'"[16] A little rhyme of 1914 expressed the importance of eggs for the moderately well-off small farmer:

> Cackle here and cackle there,
> Lay your eggs just anywhere;
> Every time you lay an egg,
> Down the mortgage goes a peg.
> Cackle, cackle all the day,
> Who can find a better way
> For to get ahead again
> Than to cultivate a hen?[17]

The Irish government and various agricultural groups got into the egg effort. In 1892, the Congested Districts Board that operated in poverty-stricken areas in western Ireland advanced a poultry improvement program as a business enterprise. It set about distributing birds (Black Minorca, Indian Runner and Plymouth Rock breeds) to remote areas for eggs, later distributing eggs for hatching instead. The Irish Agricultural Organisation Society, founded in 1894, sought to promote cooperation, extension and modern agricultural methods of new breeds, cleaner facilities and so on.[18] From the early twentieth century, the Department of Agriculture and Technical Instruction provided selected farmers and farmers' wives financial assistance to help them to build up their poultry business. By 1911–12, almost 400 egg distribution centers were scattered throughout Ireland, distributing 68,000 dozen eggs. It is true that the administrators of these organizations worried about the dominance of women, both in their poultry training courses and in the poultry-farming community at large. They further believed that none of the schemes would ultimately succeed unless men entered the poultry-rearing business. As a result, relations between the poultry societies and women raising poultry were tense. But the effort to substitute male for female poultry-keepers ultimately failed, and farm women carried the flag of the successful chicken farmer.[19]

When eggs began to hatch in Vineland, New Jersey, avian specialists started to arrive. The sciences, food products and fascinating people converged in Vineland from the mid nineteenth century, after

Charles K. Landis purchased over 200 square kilometers (km^2) of land in the 1860s and 1870s as a site for a temperance-based utopian society founded on agriculture and progressive thinking. Temperance Town encouraged settlers by promising inexpensive land in exchange for clearing the forest for farms, with planned, wide thoroughfares to connect the hopeful people and their industriousness. Once he determined that Vineland was well suited for growing grapes, Landis attracted Italian grape growers, offering them roughly 8 hectares of land for grape growing, but not for wine. Here Thomas Bramwell Welch founded Welch's Grape Juice; here a glass-making industry was established; here arose the Progresso Soup Company.[20] Here, also, progressive "eugenics" found a home in the scientifically suspect work of Henry Goddard, who established the term "moron," assembled the sensational (and false) *The Kallikak Family: A Study in the Heredity of Feeble-Mindedness* (1912), and whose studies "proved" that such defects as prostitution and criminality were hereditary. Here the authorities founded New Jersey's first schools for mentally infirm patients, where Goddard opened a laboratory at the Training School for Backward and Feeble-minded Children in Vineland. And here, there were eggs, millions of them. In the first half of the twentieth century, Vineland embraced the poultry industry as the "Egg Basket of America," as perhaps the Petaluma of the East, once home to the largest egg auction in the country, of hundreds of thousands of eggs daily. One veterinarian, Dr. Arthur Goldhaft, set up a poultry practice in Vineland, eventually opening the Vineland Poultry Laboratories, where he discovered a fowl pox chicken vaccine that saved millions of chickens from premature death.[21]

The upper classes were crucial in the history of the chicken, as were colonial endeavors that spread knowledge of the diversity of the animal. Queen Victoria played a major role in popularizing being a collector of chickens for their beauty to display them in aviaries. The Queen and her husband, Prince Albert, loved owning poultry and breeding chickens, and especially Cochin chickens. Queen Victoria also enjoyed eating fowl – and to eat in general, although she preferred fatty mutton and beef. She accumulated many of her fowl from the reaches of the Empire where the sun never set. Chinese chickens arrived on military and trade vessels from captains and others who wished to ingratiate themselves with the Queen – in particular, after the Opium Wars with China, and the opening of the nation through European force. According to the *Illustrated London News* for December 23, 1848, by that year Queen Victoria's considerable collection of birds occupied "a half-a-dozen very extensive yards, several small fields, and numerous feeding-houses, laying-sheds, hospitals, winter courts, &c." The new houses held the "curious" Cochin-China fowls "of gigantic size." The birds had a "rich glossy brown" color, "tail black, and on the breast a horse-shoe marking of black; the comb cleanly and neatly formed, with shallow serrations; the wattles double." The newspaper also reported on the magnificent deep mahogany color of the eggs and their wonderful flavor.[22]

The Queen's Light Brahma Chickens, also from Shanghai like the Cochins, gained public attention via the American birder, George P. Burnham, who presented some of the fowl to her to gain publicity and endorsements for his business – and indeed received a portrait from the Queen for the inside cover of his book *The China Fowl: Shanghae, Cochin, and "Brahma"* (Boston, 1874). Burnham claimed that a sea captain had transported 100 China fowl:

> of all colours, grades, and proportions. Out of this lot I selected a few grey birds that were very large, and consequently very fine. I bred these, with other grey stock I had, at once, and soon had a fine lot of birds to dispose of – to which I gave what I have always deemed their only true and appropriate title (as they came from Shanghae), to wit, Grey Shanghaes.

He distributed the birds widely and finally determined to give a few of the choicest ones to the Queen. He departed China in December 1852 and arrived in London in January 1853, with a consignment of "nine beautiful birds, two males and seven pullets." They were "prolific," weighing about 10 pounds each. The *Illustrated London News* printed a long article on the gift of Chinese fowl to the Queen along with a plate (engraving) and description of the creamy white color, splashed with light-straw highlights, and the upright tail in black. The Queen thanked Burnham for the gift by letter.[23]

Chickens have served in myth-making. Robert E. Lee, a slave owner who led the Confederate Army in battle against Union troops, a man who opposed proper reconciliation after the war, and for his efforts was denied a pardon (until pardoned in 1975 by President Gerald Ford, the same man who pardoned Richard Nixon in 1974 but failed to offer amnesty to draft dodgers), commanded great respect among his soldiers for his fearless leadership, support of slavery, and hatred of the north. Less known is that he had a chicken as a pet from early 1862 – a small black hen that escaped from a shipment of fowl sent to the Army of Northern Virginia, secreted herself in a tent and laid an egg, fortuitously in Lee's tent. "Nellie" remained in the company of Lee's army, laying an occasional egg for herself, or for him, that he dutifully ate; according to one version, she laid an egg almost every day for Lee, but this is a near-impossibility for birds of the nineteenth century. Having lost the Battle of Gettysburg, Lee and his soldiers retreated, somehow with Nellie, who survived even the Battle of the Wilderness. Lee's slave servant, William Mack Lee, ordered to prepare a dinner for Lee's generals, decided that the large table of distinguished guests required good food, found Nellie, killed, plucked and stuffed her with bread and butter and served her up. Lee apparently regretted the loss of Nellie deeply.[24]

The chicken of the pre-industrial CAFO era, the free and bounding birds of the open range, found response in fictional literature as well, where readers, including children, could infer something about the lives and behavior of chickens. Indeed, I can find no children's literature of the life of a sweet chick in a CAFO. In his *Gockel and Scratchfoot*, the mid-nineteenth-century artist and fairy-tale writer Gustav Süs, known for his animal drawings, mostly of poultry, tells a children's story about an old mill operator and his wife, and their chickens, turkeys and pigeons, and, in particular, the rooster Gockel and hen Scratchfoot. The chickens strut about after feeding; they seek solace; they hide in the shade. As for the story-line, Gockel and Scratchfoot become a pair of platonic love birds who have great and risky adventures in the mill yard and beyond, involving fights with other birds, encounters with foxes, and the experience of the miller's wife's fury. And, throughout it all, Scratchfoot and Gockel hardly sit still in enclosed, environmentally controlled buildings, but feed, jump, peck, fly, roost, crow and sing, and eventually return to the miller's yard to rule the roost.[25]

In *Chicken World* (1910), E. Boyd Smith illustrated the colorful ecology of the chicken. In a series of plates, he showed chickens laying eggs, sitting with the brood, hunting and pecking for worms and grubs with chicks, exploring for bugs, resting in the shade of plants, entering the clean and airy coop, roosters fighting over and re-establishing pecking order, and a final drawing of a chicken, plucked and baked, in a shallow dish and ready for carving.[26] In the series of "Here and Now Books," children lived "in the moment," while learning from "their own thoughts and observations." The books, written by the founder of New York's Bank Street College of Education, Lucy Sprague Mitchell, focused on the routines of everyday life. The Here and Now stories included "The Rooster and the Hens," in which Mitchell traced the life of the chick from the egg, pecking through the shell to get out, his immediate "scritch, scratch, with his little foot," quenching his thirst from a little saucer of water, and then the birth of other chicks, all looking for insects and worms, scritch, scratch. They grew quickly, some getting a great big red comb on the top of their heads and under their bill, and long spurs on their ankles, and eventually learning to scream "Cock-a-doodle-doo." In a story, told at times in rhyme, the child learned about nesting, brooding, egg laying, and clucking and crowing like birds.[27]

The British poet Walter de la Mare focused on writing for children. He offered this ditty on the excited behavior of poultry:

> Clapping her platter stood plump Bess,
> And all across the green, Came scampering in, on wing and claw,
> Chicken fat and lean: Dorking, Spaniard, Cochin China, Bantams
> sleek and small,
> Like feathers blown in a great wind, They came at Bessie's call.[28]

The American poet Robert Frost worked as a chicken farmer in Derry, New Hampshire, in the early 1900s, before turning full-time to poetry, and published 11 articles, mostly fiction, in two agricultural trade journals: the *Eastern Poultryman* and *Farm Poultry*.[29] In one of Frost's articles, "Trap Nests," Frost describes the troubles met by a couple starting a backyard poultry flock as they try to force uncooperative hens into laying by putting them into trap nests. The story indicated the challenges of poultry farming and the difficulties of managing a flock. In "The Cockerel Buying Habit," Frost offered a conversation between a poultry farmer and an unnamed narrator about the state of a flock with some suspicious genetic variations, and the possible source: a young male chicken – a cockerel – purchased from a "so and so a few years back."[30] These and other writings celebrated the New England, Puritan work ethic. One of Frost's poems on chickens, "A Blue Ribbon at Amesbury," described "such a fine pullet" with blue ribbon honors – "her golden leg, her coral comb, her fluff of plumage, white as chalk, her style, were all the fancy's talk" – and with her egg-laying prowess and the majesty with which she mounts her roost.[31]

Because of their energy, their fondness for the sun, their love of grubs and so on, it would seem that all cultures and languages use the chicken as a focus of children's books to show young people that sun and air and animals are a part of the natural world. A search on google.fr for "livres pour enfants sur les poulets" offers hundreds of results. *Comptage avec le Poulet Sam*, *Petit Poulet*, *Les Poulets guerriers*, and many other books fill the shelves of bookstores in the children's section in France. Two books caught my attention: *Le Poulet fermier*, about a chicken whisperer, Douglas, who wishes to become a farmer, but he does not know how to milk cows, believes that carrots grow in trees and can only drive the tractor in reverse. The surrounding farmers all laugh at him. Father Tropenjus, for sure, will never let him marry his daughter, the charming Miranda. Fortunately, Douglas knows how to talk to animals. That's a good thing, because in his henhouse, there's a chicken called Ernest, a wonderful chicken that helps transform his life. And I enjoyed *Histoire du grand méchant poulet*, about a time long ago when the hens had teeth and the wolves did not have any, and when, in a hut deep in the woods there lived the Grand Naughty Chicken. This hopeful book suggests that chickens forced to live in CAFOs may someday grow teeth and attack their handlers in a naughty way.

Science, Clubs and Beauty Pageants

The chicken moved from literature and cockfights, from a source of income for people marginalized by pogroms and subsistence agriculture, from being humanizers of traitorous generals and powerful queens, to a cultural phenomenon by the late nineteenth century, and to the focus of research, innovation, laboratories and clubs. As the possibility of sales in distant markets expanded during the industrial revolutions and urbanization of the nineteenth century, new pressures and possibilities arose for farms, including that of selling eggs. This meant that the chicken gained importance for many farms across the world not only as a source of supplemental income, but also as a potential marketable product. Toward this end, agricultural researchers and public officials sought to engage society in promoting chicken culture. Children's clubs and organizations advanced popular agricultural knowledge of chicken farming through activities that were usually connected with extension services. Many of these clubs and organizations were founded on notions of practical learning, public education and the inevitable progress of humanity through science. They were an extension of agronomy in existing governmental and research institutions into the public realm. To their supporters, they were a simple way to transform children into constructive members of society, and to improve rural youth who would then bring agricultural improvements, including technological ones, to their parents, the adults; farm animals were an easy vehicle for achieving that end.

In the United Kingdom, the rural National Federation of Young Farmers' Clubs enveloped local clubs from England and Wales that supported young people in agriculture. The first Young Farmers' Club opened in 1921 in Hemyock, Devon, where the United Dairies milk factory engaged children in calf-rearing competitions. The focus of clubs expanded to include calves, pigs, poultry, bees and gardens, and to some 50 clubs. In the US, those organizations with agricultural interests included boy scouting, 4-H clubs ("head, heart, hands, and health") and special poultry clubs.

The Boy Scouts of America encouraged industriousness, virtue, discipline, independence and practical knowledge by awarding merit badges after "significant achievement" in a wide variety of areas. Those in agriculture focused on beef cattle production, rabbit raising, gardening, landscaping, agribusiness and agricultural mechanics. Forestry, and Soil and Water Management, were the two most popular agriculturally related merit badges among the original 57 badges listed in the first Boy Scout *Handbook* (1911). Others were for achievement in Agriculture, Bee Farming, Dairying, First Aid to Animals, Forestry, Gardening, Horsemanship and Poultry Farming. "Poultry Farming" as a merit category existed from 1911 until roughly 1975, when it was subsumed into "animal science" generally, along with beekeeping, animal industry, beef production, dairying, hog production and sheep production.[32]

As in other fields, the agriculture merit badges required significant effort and mastery of information, along with the requirement to "grow at least an acre of corn which produces 25 per cent. better than the general average"; to "be able to identify and describe common weeds of the community and tell how best to eliminate them"; and to have "a practical knowledge of plowing, cultivating, drilling, hedging, and draining," and also "of farm machinery, haymaking, reaping, loading, and stacking." The merit badge for Poultry Farming indicated a knowledge of incubators, foster-mothers, sanitary fowl houses, and coops and runs; understanding of rearing, feeding, killing and dressing birds for market; the ability to pack birds and eggs for market; and the raising of a brood of not less than ten chickens.[33]

The spread of poultry festivals accompanied the expansion of clubs, trade fairs and other public displays. In January 1919, owners entered over 7,000 birds in a poultry show at Madison Square Garden to display their creatures.[34] Along with the later "Chicken of Tomorrow" contest (see below), the Delmarva broiler industry sponsored a "Chicken of Tomorrow Queen," Nancy McGhee of Berlin, Maryland, who, with her court, road in a carriage upholstered in white chicken feathers in a white gown.[35] But contests for chicken beauty and performance have attracted considerable attention and a large number of entries.

The USDA Cooperative Extension Services that were connected to the land-grant universities in order to inform people about current developments in agriculture, home economics, public policy and government, were established with the Smith–Lever Act of 1914. In existence through local and national initiatives before the passage of the Act, by 1915 Boys' and Girls' Poultry Clubs, in Georgia alone, had already enrolled 600 boys and girls. The hope was that extension programs, in part through the clubs, might transform Georgia agriculture from tenant farming and cotton-based into something more dynamic. As it was, poultry were of "very poor grade": Georgia scrubs that meandered over the farm and did not even have roosting facilities.[36] Dozens, and then hundreds, of clubs formed.

The 4-H program in the US, founded also in 1914, grew out of the efforts of the Cooperative Extension Service to expand club work. The clubs reflected broader social goals than promoting improvements in agricultural technology and animal husbandry, and sponsoring contests. If, at its roots, 4-H was seen as a way to introduce resistant rural farmers to new technologies, soil science, hybrid animals and seeds that were developed within land-grant universities, then 4-H was also intended to reflect social norms. Unfortunately, 4-H club practices and mottos reinforced traditional, regressive ideas about race and gender. In the first years, girls were trained in home economics, child-rearing and hostessing, and boys in raising crops and animals. By the 1950s, 4-H promoted a kind of anti-Communist, pro-agribusiness agenda. 4-H clubs were segregated, even if the citizenship oath of members promised to fight against "tyranny and injustice." For African American children, clubs were run by the separate system of black agricultural colleges and extension agents, whose agents earned less and which generally were poorly funded. After the 1954 US Supreme Court decision against segregation, *Brown* v. *Board of Education*, and then the Civil Rights Act of 1964 made segregated 4-H clubs and camps illegal, a number of the African American ones shut down for lack of resources. The "Negro" clubs were served solely by black agents, and the clubs and extension efforts were poorly funded and poorly supplied by the government in comparison with white clubs.[37] It should be apparent, even to the most obtuse racist, that chickens would be colorblind and needed clean air and water, good feed and free range, and that black and white children could equally – not separately – raise them well.

By this time, agricultural agencies worldwide promoted wholesome and safe preparation of foodstuffs in the name of public health, but also of benefit to specific industries. The USDA got into the business of promoting chicken both through informational publications and consumer pamphlets – for example, how and why to bring poultry to the dining-room table. In one 1966 publication, an earlier version of which appeared in 1951, the USDA observed that "flavorful poultry has long been a bargain in good eating. Now it is even more popular because it is available in convenient sizes – chilled or frozen – the year around." The department touted the variety of prepared convenience foods; the mild flavor of chicken; the good nutrition as an excellent source of protein, iron, thiamine, riboflavin and niacin, and the relatively low fat content. Chickens could be roasted, broiled, fried, rotisseried, barbecued, baked, simmered, steamed, braised and pressure-cooked, and those sold as US Grade A were nutritious, wholesome and safe.[38]

Chickens, Consumers and War

Poultry farming came into its own as an independent focus of meat production in the 1920s, and became not so much a source of supplemental income or an item for local markets as a commodity for distant markets. At this time, its treatment as a cultural icon became subsumed into a productivist and consumerist ethos. That is, the chicken was no longer a vibrant, scattering, singing member of the farmyard, but an egg layer or piece of meat, and hence its physical attributes would be reworked literally and figuratively. The US seems to have led the way in changing the chicken into an industrial object and meat bird, followed by Brazil and China and others.

One impetus to industrialize chicken has been the world wars of the twentieth century that led governments to ramp up production of armaments, ships, tanks, airplanes and the like; to engage in requisitioning of millions of men into armies as soldiers; and to control – or at least plan, encourage and closely monitor – food production to send the sustenance, itself now a military product, off to battle. Yet feeding millions of soldiers abroad meant rationing of some foods – beef and grain – at home. The resulting shortages of those foods meant stimulus in the production of others for domestic markets. In the US, World Wars I and II led to the expansion of the poultry industry; of course, it helped that fighting never occurred on US soil to devastate land, facilities, animals and people, and this enabled a turn to other products. During World War I, officials created the National War Emergency Poultry Federation to protect and develop the industry and ensure a continued level of profitability during potential disruptions.[39]

During World War II, the poultry industry benefitted again from rationing of beef and pork in order to feed the millions of soldiers and sailors in the war effort. Chicken and fish replaced beef for many consumers, and the industry thrived until a difficult period of adjustment after the war when beef and pork became available again. After the end of the war, even with many products again in stores with the end of rationing, some products remained in short supply – among them grain and wheat needed for overseas shipments. Government radio program broadcasters in 1946 therefore urged consumers to turn to eating eggs, frying-chickens and canned citrus products, and in some regions of the country encouraged them to eat more potatoes and citrus. They noted that the "poultry picture is excellent right now," with several times more broilers and turkeys in cold storage than a year before, and buying more fowl "will take pressure off livestock supply" by enabling the feeding of livestock at very high rate.[40]

Contributing to the rise of the industrial chicken was the replacement of the butcher by the supermarket. What came first, the chicken or the supermarket? In the US, tied to the emergence of the supermarket in the 1930s as the place where more and more people shopped, the perfect bird would be bigger, rounder and with meatier breasts, and would come wrapped in plastic and bloodless. Already in the 1930s, ready-packaged, cut-up, frozen and easy-to-prepare foods had made their way into display cases. Large-volume production-line methods were capable of cutting and wrapping the birds to prepare them for long-distance shipping, while improved sanitary procedures ensured some confidence in the wholesomeness of products. Supermarket display cases were bulging with ready-to-cook, precooked and frozen poultry.

This situation – the end of war and the rise of the supermarket with its strong interest in promoting such products as chicken meat – brought economics fully to bear on chicken culture, especially after one business nearly single-handedly contributed to the rise of genetically selected meat birds. The nation's largest supermarket chain, Great Atlantic & Pacific Tea Company (A&P), carried out a campaign to convince the nation's taste buds to relish poultry. A&P Supermarket was for foods, by the end of the 1920s, what Walmart became by the end of the 1990s for retail in the US. Its "Chicken of Tomorrow" Program was directed to speed up the practice of industrial raising of commercial meat chickens on a mass-production scale, and to capture consumer interest. The program seems to have originated in an address by Howard Pierce, Poultry Research Director at A&P, who remarked what a boon it would be to breed chicken like a turkey, with more meat on the breast, thigh and drumsticks that the consumer would enjoy. This led A&P to sponsor and underwrite the cost of a long-range program to improve meat-type chickens. A&P sought to identify precisely such a broiler, with more meat, a high feed-to-weight conversion, more breast meat in particular, but also "bigger drumsticks, juicier thighs, and layers of white meat." A&P hoped that farmers, who tended to egg production and ate their laying hens on a special occasion after their laying utility had ended, would consider the production of meat chickens full time.[41]

The chicken interests in the contest worked through an industry committee with representatives from ten national poultry organizations, two journals and the USDA. The committee appointed state Chicken-of-Tomorrow programs where there was interest. The committee held contests on a three-year cycle: the first in Georgetown, Delaware, in 1948, and the second at the Fayetteville, Arkansas, experimental station, in 1951. In both, chicks were raised in controlled conditions, were then evaluated on an economy-of-production score card for eggs, and a dressed-chicken score card for meat, whose criteria included: a well-proportioned body; a broad breast, long and full "meated"; thigh joint and drumstick, also full-meated; and the carcass free from pinfeathers, skin bright, soft, pliable and smooth. The contest involved some 40 finalists, each of whom submitted 18 eggs to a central facility to be hatched and raised in controlled conditions over 12 weeks to slaughter, to measure weight, health and appearance.[42] In 1948, and again in 1951, Arbor Acres farm in North Carolina won in the contest's purebred category with its Cornish – White Rock mix, although Red Cornish crosses from the Vantress Hatchery outperformed them. Eventually, the breeds were crossed to become the Arbor Acre breed whose genetics now dominate poultry farms worldwide.[43] As one example, over half the chickens raised in China have a genetic link to the Arbor Acres stock.

Driven by processors' demands for more and cheaper chicken, the broiler industry went through a rapid vertical integration process with hatcheries, growers, feed mills and processors all merged into larger and larger commercial farms. The owners of Arbor Acres, the Saglio family, helped found the National Chicken Council, the chicken industry's public relations and lobbying organization. Nelson Rockefeller bought the company, which is now a global power. The losers of the

Chicken of Tomorrow contest are gone, as are their breeds, with small farms and genetic diversity giving way to agribusinesses and standard chicken units.[44] In 2008, William Muir et al. published a study that concluded that half of the diversity of ancestral breeds had been lost. As with all monocultures, this made chickens more susceptible to diseases and blights.[45] Corner stores, also, have departed the scene in many places, to be replaced by massive supermarkets, and fewer and fewer of them sell chicken meat, although hot dogs and sausages abound, while the happy winning chickens of the 1948 contest now spend their time alive in a fowl-smelling, overcrowded cage, fattening twice as much in seven weeks, half their former lifespan.

Chickens remain an object of awe and celebration, even if their major function is to be meat and eggs. Every state has at least one poultry show (Alaska, for example), and many have several, usually connected with a state or county fair. Poultry fairs extend far and wide, across Europe to Asia, from Africa to South America. The Deutsche Junggeflügelschau Hannover offers "the possibility of buying and breeding accessories and fine breeding animals on the German poultry stock market." It includes a competition for the most beautiful chicken, and provides "a framework for the whole family." Among the larger producers, the events are annual. In nations with a smaller industry, they may occur once every three or four years. The events are apparently heavily attended and provide access to new knowledge about breeds, feeds, other supplies, displays and technologies of production.[46] Many of these trade shows and fairs date back to the early part of the twentieth century.

Brazil has established the "Capital of Eggs" in Bastos, in São Paulo state, with a population of 21,000.[47] The town was settled at one time by Japanese immigrants, and, having tried coffee and cotton production, the city fathers laid eggs, lots of them. Bastos has the largest flock of laying hens in the country, the highest production of eggs in the nation, and is the site of the annual Egg Festival, usually in July. The Egg Festival was born in 1948, when the local Japanese colony celebrated the twentieth anniversary of Bastos. Initially known as *Nyushokussai* because of the relationship with the strong Japanese culture present at the time, the event was held by the Japanese Association and involved such cultural attractions as the Ikebana exhibition (floral arrangements), dolls and dances. Originally held every five years, as the chicken became an international commodity crucial to Brazilian agriculture, so the festival was transformed into an annual event that offers gastronomic delights, craft exhibitions, folk dances and contests. The Festa do Ovo celebrates a partnership with the Rural Union, Bastos Town Hall and the Nikkey Sports and Cultural Association (ACENBA), and includes an exposition of horticultural products, agricultural implements and machinery, vehicles, computers and, of course, poultry products. It has grown massively; in 2015, 120,000 people attended the three-day festival held in the 40,000-square-meter Exhibition Center Kisuke Watanabe. The competitive 2018 Egg Quality Contest drew a jury of 18 judges with poultry expertise, including Dr. Carlos Aranguren, a veterinarian who represented the company NOVOGEN that uses genetic engineering to modernize the egg as a package of protein. Thirty-seven ova competed in the brown egg category with Granja Ueyama winning the first-place trophy – perhaps with a standing ovation. The Granja Ueyama farm had won prizes in previous years.[48]

Beauty contests continued into the twenty-first century with several notable tournaments and displays. In Malaysia, the Ayam Serama (*ayam cantik* = pretty chicken, or *ayam katik* = pygmy chicken), a true bantam-class chicken – indeed, the smallest chicken in the world – has drawn significant attention in books, shows and contests. Also known as the brave warrior, toy soldier or archangel chicken, the pygmy bird was first exhibited in 1990. It has recovered from a 2004 Avian Influenza (referred to here as Avian flu, bird flu and so on) epidemic, to capture public fascination. Weighing 340–570 grams, a full-grown rooster is the size of a pigeon, and, according to Malaysian folklore, the chance cross-breeding between a pigeon and a chicken resulted in the tiny bird. Further efforts to miniaturize the pygmy chicken have produced 180-gram cocks and 150-gram hens, and led to sale and collection as household pets, especially considering their friendly and calm demeanor. In Malaysia, they are more popular than cats and dogs. While of no standard color, the birds have distinguished red wattles. Wee Yean Een from Kelantan, Malaysia, named and probably established the breed. The Kapan, Chinese Silkie Bantam and the Japanese Bantam were used in developing the breed, but the short-legged influence of the Japanese Bantam can lead to a lethal short-leg gene which makes a chick unable to move, and thus unable to peck out of the egg to hatch. Each weekend throughout Malaysia at various chicken shows, hundreds of entries compete to be the most beautiful Ayam Serama. The chickens have reached the US and UK, although they prefer warm to cold climates. Photographer Ernest Goh captured the strutting style of the chickens, which have been "groomed to be beautiful, majestic, and strong,"[49] in his photography book, *Chickens* (2015).

In Vietnam, a country with 250 million chickens and a burgeoning broiler industry, the celebration of fowl led to national beauty contests. Vietnam's poultry industry has a distinguished history, with chicken husbandry beginning some 3,000 to 3,500 years ago in the Tam Dao and Ba Vi mountain valley areas, according to archeological evidence. At present, poultry production occurs throughout the nation, but is concentrated in the Red River and Mekong River Deltas, the northeast and the southern part of the northeast region, and is the second most important sector of animal husbandry behind pork, although only making up 15–17 percent of the industry.[50] Recently, the Dong Tao or dragon chicken has drawn attention, with the first national beauty contest in 2017 in Dong Tao commune, 30 km from Hanoi, to celebrate the massive bird, which can weigh up to 6 kilograms, with huge legs as thick as a wrist, that resemble reddish, gnarled tree trunks. Once bred exclusively for the country's royal families, today a beautiful dragon can sell for VND40 million ($1,760),[51] a huge sum given the average annual GDP per capita of $2,400.

In October 2013, to celebrate both beauty and the owners' wealth, a Saudi farmer invited other Gulf state birders for a chicken beauty contest that attracted scores of chickens and their owners. Mutaab Al-Othman held the contest at his farm near the capital Riyadh. Some chickens might generate SR2,000–10,000 ($600–3,000) at the contest auction; at a previous auction, several breeds sold for more than SR35,000 ($10,500). The cliché is that beauty is in the eye of the beholder, and contest criteria were unclear. But the contest was in keeping with other Saudi beauty contests for horses, camels, goats and falcons.[52]

3 *The industrial, as opposed to religious, slaughter of chicken in a modern processing facility, São Jose Santa Catarina, Brazil.*

Believing in Chickens

Many people glorify and venerate chickens – worship them, if you will – in different ways from those connected with clubs, business competitions to identify advantageous breeding qualities, and beauty contests. They are people who engage in a variety of celebratory, introspective and other practices connected with religious beliefs and rituals. If moral concerns about animals had greater meaning, then the world's great religious traditions might be expected to be at the forefront of efforts to encourage their followers to treat animals with a degree of sensitivity or respect.

Animals are part of god's glorious creation, and, indeed, various texts encourage a respectful attitude toward nature. Nuns, imams, priests, monks and rabbis may speak about kindness to animals – and perhaps to their fellow humans – yet rarely urge people to give up their animal protein. Monks, for example, were to earn their keep and use the surplus – in the sale of honey, beer and cheese – to help those in misfortune. Around the world, people in religious orders make bread, olive oil, nut paste and, again, beer. The preparation and sharing of meals have been an essential aspect of many religions, as has the celebration of godliness in festivals of eating and fasting. Religious figures may encourage killing of animals according to holy laws and traditional practices, or avoidance of some kinds of animal proteins in various practices (kosher, halal and so on). The point here is that guidance on what to do about the conundrum of the efficient killing nature of the CAFO has not come from, and likely will not originate in, religious traditions. The broiler chicken faces Jewish, Muslim and other culinary fates.

People worship the CAFO chicken at the industrial altar; it lacks any human meaning beyond its use for consumption. In the religious world, chickens help people deal with questions of life, morality, hope and immortality. In religious ceremonies, chickens have been near the top of the ritual sacrifice order. Orthodox Jews and Muslims slaughter them, and other animals, in meaningful ceremonies. Let us remember that the Israelites engaged – and engage – in ritual slaughter, with blood sacrifice enabling forgiveness by god for their sins, while the blood of Christ, at least symbolically, also redeems believers from sin. In Leviticus 9:3-4, the scriptures urge sacrifice:

> Then to the sons of Israel you shall speak, saying, 'Take a male goat for a sin offering, and a calf and a lamb, both one year old, without defect, for a burnt offering, and an ox and a ram for peace offerings, to sacrifice before the Lord, and a grain offering mixed with oil; for today the Lord will appear to you.'

Many African people kill chickens and goats – their blood a means of communicating with gods and relatives. At the entrance to Tongnaab Yaane, the shrine and dwelling place of the West African deity Tongnaab (Chief of the Earth), "rotting carcasses sit piled atop cement pillars, and feathers and bones stick to vertically standing logs. Dried blood covers everything." The shrine, one of hundreds in Ghana's Tongo Hills, celebrates animal sacrifice – chickens, goats, sheep, donkeys, cows and dogs – in the search for "fertility, stability, prosperity, and security in life."[53]

While many rural African traditionalists practice Christianity, they still rely on healers and ritual specialists to mediate between them and the ancestors. Many of the mediators invoke the powers of ancestors through various dances and special dress and skin painting. In the Bantwane community, the installation of ritual specialists occurs in major ceremonies that involve many candidates – and animal slaughter. At the initiation ceremonies, novice ritual specialists consume gall from ritually slaughtered goats. A chicken is also commonly placed on the head of a prospective candidate before being slaughtered, and the initiate is then covered in the chicken's blood.[54]

Orthodox Jews have for centuries performed Kaporos, a ritual that brings atonement and redemption. The slaughter creates cognitive dissonance for some people in the twenty-first century since it is carried out in public – for example, in some New York City neighborhoods. Since Kaporos occurs on the open streets, many people who reside and work in, or travel through, Brooklyn neighborhoods encounter it, and some have found it disturbing. Kaporos involves grasping a live chicken and swinging the bird three times overhead while saying a prayer that symbolically asks God to transfer sins to the birds. The rabbi or other religious leader then slits the chicken's throat – each year, thousands of chicken throats. The chicken bleeds out, is considered kosher, and its meat is to be donated to the poor and others in the community. In New York, the practice disturbed members of the Alliance to End Chickens as Kaporos, and also passers-by, who filed suits against the ritual chicken killers to end what they claimed was a health hazard and cruel, and they were disgusted by "dead chickens, half dead chickens, chicken blood, chicken feathers, chicken urine, and chicken feces on the streets." They also sued, as defendants, the New York City Police Department for failing to fight the fowl mess and the non-City Orthodox Jewish rabbis who arrived in Brooklyn from *shtetls* in upstate New York, to carry out the kosher killing. But recognizing the "constitutional issues implicated by governmental involvement in religious activities," the New York Supreme Court determined that there were no laws to decide whether to prohibit – or permit – Kaporos, and that the Plaintiffs lacked clear legal right to dictate which laws were enforced, how, or against whom.[55]

Similarly, many animal rights activists also worry that Muslims who kill animals, even if – or especially if – according to halal rules, subject those animals to cruelty and suffering. Halal rules seem to prevent stunning before slaughter, before cutting the throat of the animal.

A number of NGOs in several countries, whose members are worried about animal welfare, have asked for religious slaughter without stunning to be banned. And while it seems most animals, even under these rules, are slaughtered after having first been stunned, still, in the case of Islam, with Muslims one-quarter of the world's population by 2030, what would halal slaughter rules mean for animal rights? EU laws, for example, allow exceptions for religious slaughter, although in Sweden religious slaughter has been forbidden since 1937.[56] In any event, permitting the practice of ritual industrial slaughter of millions of chickens by conveyor belt, perhaps stunned first, hung upside down, throat slit, bleeding out – but out of public sight – at CAFOs and other large meat operations, while opposing Ghanaian, Jewish or Muslim practices of chicken slaughter, is likely inconsistent, at best, from a moral standpoint. To put it more directly, these past practices and attitudes connected with religious slaughter are also reprehensible, and it is best for humans to adopt a new approach to getting or making food, and toward the goal of limiting our anthropocentrism.

Indeed, this book considers another kind of slaughter ritual, industrial slaughter of broilers in factory farms. How quickly things have changed since the mid twentieth century, when chickens, largely, were raised and slaughtered at the small family farm – or even if in the first integrated facilities, they seem to have been killed according to more humane standards, or at least on a smaller scale. The industrial imperative to save energy and time at each step of the production process, to find a standardized, mass-produced chicken unit, to minimize inputs of light and feed, and to slaughter, cut up and distribute as quickly as possible, without any sense that the bird was a creature and a part of human culture, had not taken hold. But, as the next chapters point out, the only religion in the factory farm is profit, and past understandings of chicken biology, ecology and behavior have been shoveled out of the shed to be dumped on the manure heap of history.

2

Ecology and Industry

> Science knows no country, because knowledge belongs to humanity, and it is the torch that illuminates the world.
>
> – Louis Pasteur

What is "chicken-ness"? What do chickens think when they hunt and peck? What does a hen feel when a fox charges toward her? How do roosters establish a dominant position in the barnyard? When chickens have been bred and selected and confined in massive industrial sheds in preparation for their space-age push to slaughter weight, how do they comport themselves? These are not questions of faith, such as those associated with religious beliefs. The answers to them have evolved over the last century. We cannot answer these questions or similar ones with certainty. But it is certain that human understandings of chicken ecology – how chickens live with other chickens, what they eat, how often they eat, what is natural and good for them to eat, how much light they thrive in, what minimum and maximum temperatures they prefer, and so on – have changed significantly. To put it bluntly, if the understandings of the ecology of chicken life that prevailed in 1900 were still prevalent, then it is less likely that intensive factory farming of broilers in CAFOs or other facilities would have come to dominate the chicken world. Today's broiler is only partly a bird. It is, in addition,

4 *Before the rise of factory farming, chickens ruled the roost – and had free range to engage in hunting, pecking and resting.*

an output of capitalist industry, created from inputs of chicken, science, energy, feed, antibiotics, light and other such things.

As a result of rapid changes in technological possibilities in the agricultural world over the past century, the closer and closer connection between industry and science, and the efforts of agriculturalists – who often work in multinational corporations – to meet growing demand among hungry consumers for protein, animal producers have turned increasingly to raising meat animals – in our case, chickens – in tightly packed, closely regulated, artificial environments like cages and batteries, and sheds. The changes reflect different understandings of chicken behavior from those only a few decades earlier, and suggest that, under the influence of politics and economics, scientific concepts and understandings change as well – for example, understandings about what is "natural" for a chicken.

Agronomists, ecologists, zoologists and animal behaviorists who today facilitate these industrial practices ignore well-documented chicken behavioral studies and human–chicken interactions over the centuries, including those of earlier ecologists, zoologists and animal behaviorists, in order to develop and support industrial production in factory farms. This literature indicates clearly that chickens are social animals, that they thrive as animals when permitted in free runs, and that they are acutely sensitive birds who think, or likely think, and surely feel pain and express their anguish. Yet in the effort to accelerate fattening from the factory to the consumer, agro-industrial specialists have used hybridization genetics and other techniques to force chickens to serve one purpose: that of growing quickly and transforming feed into meat, and they have changed the chicken habitat into an industrial setting. The ecology of chickens has also changed from the point of view of diseases: industrially raised chickens, like any monoculture of any other industry flora or fauna, are subject to new disease epidemics – for example, Avian Influenza (AI), which can decimate thousands of creatures in days, including those who must be "culled" (destroyed, killed) before they spread the disease further. This chapter explores the domestication of the chicken with a focus on scientific studies on chicken biology, behavior and habitat, and it raises the worry that chickens in factory farms are reared with inadequate attention to this rich scientific literature. Instead, studies that undergird factory farms must necessarily consider the chicken as a meat machine, and less so as a bird with a long history of domestication.

Domestication of Chickens

Domesticated animals of any sort are closely associated with humans, and rely on them for food, shelter and dispersal, whether distributed during colonization far and wide – for example, by Europeans to the New World[1] – or by farmers pushing back the forest in Asia, Europe and South America to establish agriculture, or by today's global traders. Domesticated animals reveal trade and cultural contacts between societies and civilizations. When did the chicken begin its hunting and pecking as a domesticated bird? It appeared in a variety of places in Asia, Africa and Europe millennia ago. As documented in a number of excellent sources, the domestic chicken is a descendant of the red jungle fowl (*Gallus gallus*) from Southeast Asia. The red jungle fowl inhabited field edges, groves and scrubland. Archeological evidence suggests that domesticated chickens existed in China 8,000 years ago. They spread to the Middle East and Africa, and later to Western Europe, perhaps by way of Russia. Domestication may have occurred separately in India, or also with domesticated birds introduced from Southeast Asia.

How the chicken came to West Asia, the Mediterranean and Europe after initial domestication is unclear, as is when it changed from primarily a religious and ritual creature into an economic one. Based on archeological, historical and iconographic evidence, it appears that one of the earliest areas of chicken husbandry is the Hellenistic (fourth–second centuries BCE) site of Maresha, Israel. The red jungle fowl began its trek to the West no later than the third millennium BCE, as demonstrated by chicken remains in Iran, Anatolia and Syria. There is some indication they arrived in Egypt even earlier. But, in these regions, their presence was mostly as cocks and exotic birds used for cockfighting and in royal zoos.[2]

Knowledge of the history of domestication of chickens has grown more extensive, if the story has become more complex. One recent study indicates that chicken husbandry was already present in the Yellow River basin shortly after the onset of the Holocene period. The researchers concluded that chicken domestication and husbandry in the region may have been spurred by agricultural innovations in the lower Yellow River basin, including millet cultivation, pig husbandry and dog breeding. They also note that the dispersal of poultry farming from East Asia to Asia Minor and Europe might date to the Neolithic along ancient trade routes across Central Asia rather than, as previously thought, via South Asia and Mesopotamia.[3]

The story of domestication in settings where now the chicken is a familiar bird on the farm or in stores is also more complex than usually explained. Pitt et al. have documented how the chicken, introduced into Europe during the Bronze and Iron Ages as an exotic, non-native species, may have encountered poor to limited environmental conditions for spreading, depending on the climate at the times. They conclude that "human intervention played a vital contribution during early domestication to ensure the future widespread success of the chicken," and that the first areas of domestication may not have been China. Rather, a dispersal route into Europe via the Mediterranean – with significant human investment of time and resources – may be the most likely scenario of domestication.[4]

The ancient Egyptians ate domesticated fowl and wild birds. In cities, there were poultry merchants who raised birds and fattened them to sell. The most popular game bird was goose; game hunters, and also kings, princes and aristocrats hunted quail, swans, sparrows, storks and pigeons, often using boomerangs. Geese and ducks were also domesticated, and served grilled or boiled for the banquets of kings and priests and people of status in the community. Egyptians ate domesticated pigeons, quail and ostrich. Pheasant, duck and seabirds were often salted and preserved.[5] A contemporary Egyptian cookbook highlights these ancient roots in its various recipes. It may be that chickens were introduced into Africa several times for separate needs and purposes, and dispersed in different and complex ways.[6]

Wherever they have lived, chickens have been close associates of humans – that is, until placed into CAFOs. Chickens actually appeared in the New World well before its discovery and conquest by Europeans, on boats that crossed the southern Pacific to present-day Chile, from Polynesia.[7] And today, of course, America and Brazil, North and South America, are the world capitals of chicken farming. Domestic chickens appeared in Africa many centuries ago, and are now an established part of African life. On the African continent, free-range, scavenging village chickens have long lived in all agroecological zones, from villages in the tropical rain forests of West and Central Africa, to the temperate highlands of East Africa, and to the Sahel and Kalahari deserts. They gained a hold in society in cockfights, for use as sacrificial animals, and likely only later as a new source of food.[8] They are the most abundant livestock species in Africa, with a total population of 1.6 billion in 2010, a significant change since their introduction from Asia millennia earlier.[9] Village poultry can be found in all developing countries and continue to play a vital role in rural households. They are particularly important among more remote and poorer ethnic groups, and not only in Africa but, for example, in the highlands of Vietnam.[10]

Well-Behaved Chickens

In all of these places, from Southeast Asia to the New World, from the African Savannah to Northern Europe, people have learned that chickens are social animals – at least, those that are not confined in dark, ammonia-infused CAFOs. Chickens create a pecking order or hierarchy, but find solitary time as well, and roosters tend to live independently, while nothing of the sort occurs in CAFOs. Social maturity occurs at approximately 1 year of age, although today the vast majority of chickens are slaughtered well before that time, and therefore specialists cannot be fully certain about their dispositions and learning proclivities when they are primarily objects of industry. But a close reading of agricultural journals and monographs from the late nineteenth and early twentieth centuries confirms the great advantages of fowl being raised in an environment of ample sunlight, fresh air and good food. They must have free space to exercise, develop their appetite and establish their pecking order.

In 1904, the parents of 10-year-old Thorleif Schjelderup-Ebbe placed him in charge of caring for his family's flock of chickens in Kristiania (Oslo), Norway. As a child, he loved nature, kept a journal and was particularly interested in chicken behavior. Since chicken were first domesticated, peasants and farmers had noticed that their flocks kept a specific order. At feeding time, the dominant birds in the flock "would eat first, picking out the best morsels," followed by more submissive birds, and then the least dominant pecked whatever was left. The farmers also knew that, if anything happened to disrupt this order, "introducing a new bird to the flock or removing one of the dominant birds, [then] there would be a brief period of discord as birds fought with each other to re-establish dominance," after which "peace would reign once again." Schjelderup-Ebbe explained this process: how order was established and maintained; the nature of the hierarchy; how the dominant ones pecked into submission others who deigned to infringe on food or place. There was a clear rank down the line, as Schjelderup-Ebbe noted in his earliest notebooks. Schjelderup-Ebbe studied Zoology at the University of Oslo, and wrote his Ph.D. thesis at Greifswald University: "Gallus domesticus in seinem täglichen Leben" (1921) on the social structures – "pecking order" – of birds. Publishing and researching heavily over the next decades, he continued to document how the distinct personality of each bird (and other creatures) varied from that of any other of its species, and also "a definite order of precedence or social distinction, founded on certain conditions of despotism."[11]

Schjelderup-Ebbe was joined by dozens of other researchers in the effort to improve understanding of chicken behavior, physiology and economy. They sought to give farmers the benefit of scientific studies that emanated from ever-expanding agricultural extension services around the globe, to bring practical knowledge to the masses. The first concerns of agriculturalists, however, were with the cost and efficiency of egg raising, and only occasionally with meat production. In the US, agricultural researchers touted eggs and market poultry as always ready for sale and thus a supplement to income. According to a West Virginia study, the flock "not only pays the grocery bills, but provides a considerable amount of ready money." At the time, flocks on the average farm were significantly smaller in size than in factory farms only 30 years later; they could be managed without the constant vigilance of industrial enterprises; birds lived a more natural life; and they fattened slowly. West Virginia flocks in 1915 averaged 37 fowls, but one extension worker asserted that 100 could be kept at practically no greater expenditure for labor, buildings and the like. The service offered advice on outbuildings, feed, water and how to identify and treat maladies. For head lice, for example, one pamphlet called for "greasing" the head with a small amount of lard – but not too much.[12] Good care would result in a doubling of the weight of chicks in six days, and then another doubling in six months, but there was a need for liberal feeding, cracked grain and wheat or corn, with phosphorus and lime for bones, protein for the muscular system and feathers, carbohydrates and fat for energy and heat; skim milk (sour) – plus free-range access to bugs and worms. "Exercise combined with liberal feeding" was the key.[13] Researchers concluded that whole grains were better than ground;[14] that well-ventilated, south-facing poultry houses, with slopes away from the house to ensure dryness, were crucial; that two runs were needed for each house; and, in addition to grains, a liberal supply of green food for laying was required.[15] This kind of diverse diet and regular exercise are absent in contemporary industrial aviculture.

Decades before Arkansans would live in a state known for broilers – many of them produced by Tyson, the world's largest producer of frozen and fresh chicken parts, nuggets, snacks, breaded bird, and many other products – and for Walmart, the world's largest retail store, chickens lived in a state known for eggs, not meat, and farmers and specialists knew of the need to keep their flocks healthy with chicken runs, fresh air and exercise, not the enclosed, tightly packed sheds of the 21st-century Arkansas countryside, with its 2,400 CAFOs and 108,000 employees in total. A 1908 Arkansas agricultural extension pamphlet, written in response to "constant demand throughout the state from amateurs and others, in the poultry business, for information pertaining to the raising, care and management of poultry," noted the doubling each ten years of egg production, from 1879 to 1899, with production starting to approach cotton, corn and dairy in value. In the present CAFO era, questions of location and land quality matter little because earth-moving machines can sculpt land for factory structures illuminated artificially, and cheap labor may be the crucial determinant for location. In 1908, the Arkansas extension service advised chicken farmers to seek out high land, dry, porous soil, or gravelly land that drained well – in any case, land that sloped to the south or southeast for ample sunlight. It called for orchards for poultry runs for shade during hot summers, where the birds will find many "injurious insects," and coops that permitted "full circulation of fresh air," yet were not drafty.[16]

Crucial, too, were the proper amount and variety of food, clean fresh water and exercise. Corn and grains, oats, wheat and bran should be augmented with clover and alfalfa. If free-range, the chickens feasted on insects, and specialists recommended giving them also beef or other scraps, with grit (clam shell, gravel, even plaster) to assist digestion. Indeed, "filth, dampness, improper ventilation, improper feeding and the introduction of infected birds into the yard are the most common causes of disease in the poultry yard."[17] In CAFOs, exercise and variety in diet run against the imperative to fatten the fowl as quickly as possible, and filth and dampness are the operating standards.

From the 1910s, US extension service researchers published a series of pamphlets on breeding, egg production, chicken behavior, brooding, housing and so on. No one doubted the importance of sun, fresh air and free space to achieve good results. The advice of one pamphlet instructed farmers to "pick [chickens] for appearance, a shape suitable for carrying an abundance of meat: broad and deep of body, . . . good length of back and keel, and especially a broad breast well covered in meat."[18] The author urged, when possible, to provide free range for breeding stock, leading to more exercise, better health, more fertility and more vigor. He finally advised proper, draft-free housing and sufficient space per bird: a well-ventilated and dry space, with 3 to 4 square feet (roughly 0.3 to 0.4 square meters) of floor space per bird.[19]

Throughout the world, industrial pressures for food production slowly penetrated egg and meat husbandry. In Denmark, egg production took off. By the mid-1910s, egg production had become an important industry that relied on small farmers. Nearly every farmer kept hens, especially small farmers. There were 12 million hens in 1909. The nation imported cheap Russian eggs for the home market, while nearly all of the eggs for the export market that began in the 1860s went to England – some 20 times more eggs than imported from Russia. Such a healthy export business was helped along by poultry societies that received government support dedicated to improving breeding for such established breeds as Minorca, Leghorn, Plymouth Rock, Orpington and the Danish Landhen. Inspections were a required part of the businesses, with birds numbered, fitted with a leg band, and recorded, to determine the best and strongest egg-layers. Cooperative egg export societies held contests and gave prizes for economical production, sometimes even showing up unannounced. By economical production, they meant, in addition, that the yards, houses and the like had to be in good and clean condition.[20]

An expanding research agenda was at the foundation of the modern chicken industry. Since its founding in 1908, the journal *Poultry Science* has considered a wide array of concerns to improve production and find solutions to a series of feed, shelter, disease, transport and other problems. Articles have covered feeding and supplements (including, early on, cottonseed meal as a substitute for soya, fish meal, crayfish dust meal, shrimp waste meal, to modern manufactured feeds), diseases and immunology, bacteria and viruses (*Salmonella*, *E. Coli*, later Avian flu and the like), antimicrobial resistance, breeding and reproduction, lighting, enzyme supplements, and changes in the industry. In the founding years, authors set forth the nature of the field, selection, reproduction, fertility and standardization of breeds; pathology; poultry-house construction; feed and the testing of various animal foods for chickens, and the growing use of corn and soy meal; what might be used as grit (seed kernels, for example); world experience; the role of extension services (since the journal was based in the US); illumination; and contests to select the best birds and eggs. By the 1920s, more and more articles focused on indoor breeding and raising, and on parasites, disease, vaccines and immunology, likely because growing concentration of production created new challenges in fighting ever-present diseases.

Agronomists hoped to use science and economics to bolster the small farmer, but in fact agricultural research has largely helped agribusinesses, through technological developments that they have used to mechanize processes to replace laborers; to push the soils, plants and animals with chemicals, fertilizers, biocides and hybridization processes that have increasingly well-documented social and environmental costs; and to dominate the shaping of public policies over diet and public health, producing foods that are high in salt, fat and sugars – and GMOs, with as yet uncertain consequences – leading to such problems among consumers as high blood pressure, heart disease, obesity and diabetes. Early on, however, the goal was to use research not only to push production, but also to save the family farm by promoting a kind of modern Jeffersonian ethos of self-sufficiency. The USDA Office of Farm Management, and later Bureau of Agricultural Economics, had divisions of Farm Population and Rural Welfare, Land Economics, Marketing and Transportation Research and so on, toward those ends.

Industry Joins Research

Some observers welcomed the nascent industrialization of poultry farming with deep enthusiasm for the machine age generally, while others worried about the collision of machine and animal. In 1903, Edith Bradley and Bertha La Motte took note of the rise of a broiler industry. They discussed the impetus to introduce "new appliances of every description both elaborate and simple . . . likely to revolutionize the poultry industry" then spreading in the US. They pointed to the Anglo-American Poultry Company and the Cyphers Incubator Company, whose products were ducks and broilers that were "picked up, plucked, and trussed, split open and broiled like a mackerel. They are very delicious if nicely cooked and served hot, as they possess much more flavor than the mature chicken." They described the Cyphers Duck and Broiler Plant, quite large for its time, at Wayland, New Jersey: "It consists of a 300-foot brooding house of the latest pattern, heated by hot water [, from which] during the past four years thousands of pounds of green ducks, broilers, and roasters have been produced and sold." The facility, unlike today's sheds, offered "free range," yet "of course this is all carried out on a truly American or 'mammoth' scale, which is quite beyond the scope of ordinary people, as an immense capital is necessarily required."[21]

Yet as soon as crate feeding and other industrial apparatus became more widespread, they caused some concern among specialists. Oscar Erf, a prolific researcher and writer, graduated from Ohio State with a BS in Agriculture, was an assistant professor in Dairy Manufacturing at the University of Illinois from 1899 to 1903 and a professor of Dairy and Animal Husbandry at Kansas State University from 1903 to 1907, before returning to Ohio State in 1907, as its first Dairy Science Professor. He touted eggs and chicken as part of a healthier American diet; he advocated eating less pork. He observed a three-fold increase in egg and chicken production from 1880 to 1900, and he likewise promoted their adoption among Kansas farmers. If, as in many cases, farmers who exclusively raised poultry had not in fact created successful businesses in Kansas, then there was hope. Erf contended that chicken farming offered great potential as it was inexpensive, relatively labor free, and possible without interfering with other farm work. The goal was to increase the production of standard-bred chickens, especially breeds for quicker fattening and more egg production. (He noted that the jury was still out on whether incubators were an improvement over hen hatching.)[22]

Erf worried that the new systems would deny fowl a healthy environment. He insisted that comfortable quarters, exercise, water, grit and variety of grain food, green or succulent food, and casein or meat foods were needed. Presaging the situation in future factory farms, he wrote:

> Exercise is as essential as food, and lack of it indicates wrong methods of rearing. The natural way for a chick to take its food is to scratch for it, taking a little at a time. If small chickens are put into a box with a bare floor and fed from a trough, they will become weak. Many will become clogged behind with the excrements accumulating on the down.[23]

He was therefore somewhat skeptical about the growing "crate-fed chicken industry." The "fattening of live stock by confinement" was an old idea, applied in France and England, deserved attention, and had been introduced by packing firms. Some plants had capacity for as many as 14,000 crates. Intended as labor-saving, the crates – at this time – were founded on principles of good ventilation and comfortable space for the chickens, and were designed to permit droppings and feed to be removed before accumulating and decomposition.[24] But the system required better management because the chickens, as received at the plant, "vary greatly in quality and health. Strict culling is necessary, and keen supervision of the feeding-room is required to detect and check outbreaks of disease." Elf noted that "even the best of feeders at times have poor results, due to epidemics of disease and bad weather conditions."[25]

In another pamphlet that discussed breed selection and breeding, incubation and brooding, houses and fixtures, egg production, caponizing, lice and mites, diseases and treatment, the authors underlined the fact that "care and feed are crucial" to the future of the industry.

The pamphlet touted, for eggs, such breeds as the Plymouth Rock, Wyandotte, Orpington and Rhode Island Red, and for meat, such breeds as Langshan, Brahma, Cochin and Cornish. They urged farmers to offer free range with some shade; well-ventilated, clean and dry houses; and a southern or southeast exposure, with sloping land to drain away from the buildings.[26] One of the co-authors of this pamphlet, Harry Lamon, was "Uncle Sam's Head poultryman," in charge of poultry research at the Department of Agriculture's Bureau of Animal Industry. Lamon grew up in northern New York State on a farm, and was breeding and exhibiting poultry at 10 years old, including at shows in Madison Square Garden. He moved on to manage farms and egg operations, including an Orpington farm. In government work, he helped breed such birds as single-comb white Leghorns and Rhode Island Reds. During World War I, he directed the government campaign for increased poultry production, gathering a staff of skilled specialists.[27] Together with Joseph Kinghorne, Lamon published extensively in the 1920s on selection, management, insect infestations and cleanliness, practical poultry production, feeds and feeding, incubation of eggs, and other topics.

Milo Hastings, whose views on the industrialization of the chicken were at once utopian and dystopian, offered an answer to these worries. Hastings was the inventor of the forced-draft chicken incubator, a nutritionist who created the health-food snack Weeniwinks made of grains and no sugar,[28] and writer. He grew up on a farm, attended the Kansas State Agricultural College, and entered its poultry husbandry program. He built a new kind of chicken house based on plans from the Maine Experiment Station; launched the first official egg-laying contest in America in 1904; and began ruminating about his forced-draft chicken incubator for incubating up to 1 million eggs simultaneously – quite a bit higher than the dozens in contemporary ones. He wrote about the egg trade and incubators, and published *The Dollar Hen* (1911) as a guide to free-range chicken farming, in which he celebrated running streams, fringes of trees, grass, fences, fodder and sheds as constituting the perfect habitat for birds. Around that time, he built a forced-draft egg hatchery in Muskogee, Oklahoma, that contained 30,000 eggs in 150 square feet, likely the largest in the world at that time. He subsequently attempted to build massive, commercial forced-draft chicken incubators, traveling to Petaluma, California, to sell his idea for a 1-million-egg machine. But none of his business ventures succeeded; nor did he achieve a hatchery for even 150,000 eggs.

Clearly, these ideas were connected with Hastings's determination to see chicken husbandry become industrialized. In a 1915 *Scientific American* article, Hastings asked: if dairy, shoes, textiles and pork had become industrial, then why not chickens? Was it that they needed close attention, or could not be raised by machinery, or required natural conditions? He answered that the example of yarded chickens showed they did not need free range, and pointed out that poultrymen were beginning to have success with indoor raising of chickens. Anticipating CAFOs a half-century later, he announced the possibility of raising 1 million chickens per acre. Indeed, he proposed a building covering an acre, 50 feet high, allowing 12 inches in height for chicks and 20 inches for hens, that could hold 250,000 hens or 2 million chicks, and machines that would handle the work, so that men need not be accommodated. With controlled temperatures, ventilation and blowers, with artificial lighting to ensure each level of chickens got some (he had raised hundreds of chicks in artificial light successfully), with lights turning on and off to simulate day and night, there was no limit. Further, the facility, clean and modern, would eliminate the chief cause for the failure of intensive poultry-rearing – soil contamination with disease, germs and parasites. The chicken factory, with no surface exposed but dry, smooth steel that was frequently cleaned and sterilized, would prevent disease. And as for sunshine and the vigorous outdoor activities of animals – running, walking, scratching, chasing, bathing, wallowing in dust, fighting – this could be mimicked by having water, food, baths some distance away from the birds.[29]

These new facilities, whatever their form or grandeur, created a series of challenges to raising meat birds, precisely because of pecking order, the need for ventilation and natural light, and other factors crucial to chicken health. Among birds given the opportunity to establish pecking order in less controlled and crowded situations, such as small groups in cages, aggression is relatively low. A small group size in the cages allows the hens to establish a stable pecking order. Where factory conditions may increase aggressive behavior, the manufacturers carry out a variety of violent surgeries on the birds to limit the effect of those behaviors in the batteries. They debeak female birds to prevent damage to other birds; the males do not lay eggs and are considered useless, so often are slaughtered at birth.[30] They feed, water and offer light according to specific regimens. They deny motion and space to accelerate fattening for slaughter. The chickens remain chickens, but scientists and industrialists have determined they should live by the human criterion of producing eggs and meat quickly. Thus, under the influence of human direction, these chickens differ in their shorter lives, physical transformations and disfigurements, their diet, and the limited range of their tolerated behaviors.

If pecking orders emerge in cages and breeder sheds, they seem not to emerge in meat chickens. Perhaps it is because the broilers that are slaughtered are immature (six–seven weeks) when they might have just begun to establish social stratification. In any event, bird housing has an impact on how birds react to other birds in what they recognize as a threat or aggression. But the ability of chickens to recognize and remember one another "is nearly impossible under commercial poultry husbandry conditions where flocks range to the tens of thousands of birds. Dim or colored lighting used in the cages can affect a chicken's ability to discriminate between birds; the manufacturers have opted for darker feeding barns to limit pecking, attacks, and cannibalism."[31] In any event, mortality, production and behavioral problems are all worse in large groups of hens. Most aggression is seen in feed trough competition.[32]

This aggression was not seen as unnatural at the dawn of the industrial age of broilers. Rather, quality, cleanliness and proper diet to ensure wholesomeness were paramount. A 1927 USDA pamphlet noted the ongoing transformation of fowl husbandry, with industrial practices securing a position. Most noticeable was the rapid development of the mammoth hatcheries, with "more and more farmers buying day-old chicks instead of incubating the eggs themselves." But the agricultural specialists saw the major worry as ensuring the buying of high-quality chicks "where the breeding flocks supplying the eggs for the hatchery are inspected carefully and culled rigidly, using as breeders only the best birds." It was important that the hatcheries be as sanitary as bird coops, under proper management, with chicks coming from eggs of uniform shape, shell color and shell texture. Proper feed was important. Sanitary conditions were crucial. And fresh air was paramount.[33] Thus, by the late 1920s, specialists knew that good chicken came from ventilated, open conditions with birds free to roam, fed well, and coming from the right stock. Within 50 years, tightly packed – even overcrowded – inhumane, conditions, with enzyme, antibiotic and other capital inputs, were in place to ignore or control what had been seen as the "natural" behavior of chickens, and began their spread from the US across the globe.

Indeed, on the eve of World War II, a member of the British Royal Society of Arts mused about the rapid changes in chicken-rearing. America had become the citadel of high-production, industrial slaughtering and packing. E. F. Armstrong noted that "most of us, even the most inveterate town dwellers, have kept chickens for a period of our lives, or at least lived somewhere where they are raised." But chickens had been forced to give up their hunting and pecking for confined spaces, regular food and transformation into a "protein producer" that lived on a "diet of grains and animal by-products." Here, they were at much greater risk for disease. But there was a machine solution: Armstrong called for slaughtering to be carried out "in a proper manner, so as to ensure maximum bleeding and feather muscle relaxation; a machine will do this so much better and more uniformly than man." Plucking involved both scalding and repeatedly reused hot wax, leading the bird ready to be packed and chilled for delivery and meeting the modern demand for "ready-to-cook" foods. This meant either individually wrapped and quick-frozen, or jointed and canned. At every step, there was a risk of contamination and disease. Armstrong concluded, "Such must be the new farming."[34]

In a way, all of the extension pamphlets warned presciently, if indirectly, against what would become CAFOs. The nascent factory farm would alter the chicken's life into some other kind of existence, largely by ignoring proven understandings of its behavior. A specialist in 1978 identified the crucial problem for modern poultry farming. In their search for size, speed and efficiency, poultry management experts had lost sight of the bird itself. They thought "in terms of tens or hundreds of thousands of birds." They began to behave more like mechanics than poultry caretakers; for them, birds were precisely machines with feathers and squawking sounds, which fought to avoid being manhandled and caught. To them, the chickens no longer had the behavioral and other traits of chickens. They overlooked, downplayed or ignored the massively overcrowded housing environments used to force chickens to transform along the assembly line from chicks to harvestable units, the stress imposed on the birds by the unnatural environment, and the chickens' traits that led them to fight or peck or hurt each other in this environment.[35]

What Came First, the Chicken or the Integrated Industry?

The broiler itself found its home in the vertically integrated factory farm, and soon in the CAFO. Likely, vertical integration of the poultry industry took its first steps in the Delmarva Peninsula in the interwar years. Delmarva, bordered by Chesapeake Bay to the west and the Atlantic Ocean to the east, had a relatively mild climate, sandy soil with good drainage, residents familiar with raising chickens, low labor costs compared to the surrounding region, and close proximity to major markets: New York, Boston, Philadelphia, Baltimore and Washington, DC. Apparently, Cecilia Steele of Oceanview, Delaware, was the first broiler entrepreneur, who, in 1923, mistakenly received 500 laying hens for an order of 50. She raised nearly 400 birds to 2 pounds, and made a huge profit on the meat. The next year, Steele ordered another 1,000, and by 1926 she was raising 10,000 broilers. Her broiler house has been preserved at the University of Delaware Agricultural Experiment Station near Georgetown, Delaware.

In 1997, Delmarva produced only about 8 percent of US broilers, although it employed 20,000 people directly and thousands of others indirectly in the industry. Yet, at one time, the area produced more than the rest of the nation combined. Within 13 years of Steele's first birds, Delmarva offered two-thirds of the nation's production, if during World War II it dropped to 44 percent. In the 1940s and 1950s, broiler owners worked to develop a local feed industry, too.[36] With such companies as Perdue Farms leading the way, they added hatcheries, with local farmers growing the chicks, followed by processing plants (dating to 1937), and then transport to supermarkets. Banks, hatcheries and feed companies were rather open to loans. There was very little turnover in residents, and they seemed to be innovative in boosting broiler production. Labor costs were fairly low, perhaps because 20 to 30 percent of the local population was black and worked as laborers and catchers, and rarely growers. The growers sought out more productive broilers fairly early: already by 1927, Delaware had 46 hatcheries. The demand for broiler hatching eggs soon outstripped demand, and this led to the import of better meat-type birds – the Barred Plymouth Rock and the Rock Red Cross. Still, the meat was expensive, and beef, pork and mutton consumption in the US was nine times that of chickens. Expansion of the industry, pushed by war, led in 1945 to the opening of the world's largest processing plant in Milford, Delaware – the Sussex Poultry Co., which dressed 100,000 chicks daily and had 400 employees. The film *Your Chicken Has Been to War* (1943) includes footage of the plant.[37]

The US southern states were important in the rise of integrated industry, with Georgia at the forefront of efforts, again because of climate and low labor costs. By 1952, Georgia was the nation's leading broiler state, having surpassed the Delmarva region with annual production of 100 million birds. By autumn 1952, between 2 and 3 million baby chicks were put under brooders every week. (It boggles the mind to consider that, a little over a half-century later, 9 billion chickens were consumed annually by Americans, and 50 billion in the world.) The transformation of Georgian animal husbandry was all the more surprising because, not long before, there were but a few fryers raised on small farms in the mountains and foothills of northeast Georgia. During World War I, some of the residents raised a few hundred fryers. After the war, one C. H. Wester of Temple, Georgia, began his chicken business, while W. C. Morgan, near Buchanan, started making money on fryers. These were hauled by horse or truck into Gainesville for sale. But in the 1930s, production took off, from 500,000 in 1935 to 3.5 million in 1940, to 29.5 million in 1945, to nearly 89 million in 1951.[38]

It may be that the first commercial grower was M. E. Murphy of Talmo, who grew his White Leghorn chickens in batteries (cages). But Jesse Jewell was the rooster of the industry, taking steps to create a vertically integrated operation. He opened a modern poultry plant in Gainesville in 1941, with a feed business and hatchery, by 1944 opening a rendering plant to utilize offal, and then becoming one of the first producers to package cut-up frozen chicken. If the demands of the military were crucial for growth during the war, then in the late 1940s consumer demand eventually picked up where government interest had disappeared, and Jewell and others were ready.[39] Locals say that Jesse Jewell made Gainesville the first poultry capital of the world and was the pioneer of vertical integration. He was also anti-union, so that he provided the direction for the modern poultry enterprise in the US: pro-chicken growth, anti-union.

Jewell was no progressive. If vertical integration meant the replacement of skilled male meat cutters – butchers – at the point of sale at the market, then integration meant cheap laborers on an assembly line handling the evisceration, parts production, and wrapping at the factory. There was plenty of cheap labor because of the mechanization of agriculture and decline of cotton that left many rural people looking for work – and at the mercy of employers, with low wages for dangerous work. Indeed, small farmers sought alternatives to cotton as sharecropping faded, as historian Carl Weinberg notes. In the early 1950s, a majority of workers at J. D. Jewell voted to unionize under the Amalgamated Meat Cutters and Butcher Workmen. Management apparently ordered and participated in violent attacks on union representatives, putting the effort down. Jewell's achievements were recognized as the founder and first president of the National Broiler Council, the president of the Southeastern Poultry and Egg Association, and as US delegate to the 1951 World Poultry Congress. In the early 1960s, Jewell sold his company to a group of investors, and it went bankrupt in 1972.[40]

Other than because of the capable hands of Jewell and others like him, why did the poultry industry take off in north Georgia? One answer, we are told, was good land, good climate, with foothills cool even in summer, and the fact that "many of the natives are of pure Anglo-Saxon stock – proud, hard-working, honest folk."[41] The key man in the development of broilers was the feed dealer, who, like Jewell, provided credit to the growers, and also baby chicks, feed and medicine, found markets, and helped with catching, weighing and loading the birds. The birds were raised in houses 4–5.2 meters in width, but 8 meters has become standard, in part on the basis of plans from the College of Agriculture at the University of Georgia. They could hold 3,000 broilers or more – and some even 22,000 – at a rate of about 1 per 0.07 square meters. Most houses already used automatic waterers and feeders, central heating and the like. Establishing the norm for CAFOs,

5 *As in simpler times, in less-pressured places chickens have continued to provide the occasional egg and eventually food for the table – here, with interloping duck, at Hershberger's Farm and Bakery, near the town of Berlin in central Ohio's Amish country.*

contract labor had become fairly standard, with the laborer paid not for his time, but by the pound of flesh – a few pennies for the pound.[42] All of this indicated the establishment of the vertical integration of the industry by the 1960s – by which time, in fact, 90 percent of chicken meat came from integrated operations.

By the early 1970s in the US, the significant changes in production, processing and marketing of poultry marked a radical transformation of the broiler industry, "from one of small, widely scattered farms to one that is large, concentrated, and efficient. More than 99 percent of all broilers produced are grown under contract and by integrated firms." The industry geographically was concentrated in ten states, and had linked together fully all aspects of production, marketing – and ownership. Six of ten farms raised 100,000 birds or more annually.[43] If, in 1934, annual production was large at 34 million birds, then by 1976 it had grown a hundredfold to 3.2 billion, and annual per capita consumption of broilers had increased eightyfold to 40.4 pounds. At the same time, the number of firms processing broilers dropped, so the number of birds being handled per plant skyrocketed, with, shockingly, almost 100 percent of it under voluntary health and safety inspection.[44]

The next stage in the development of factory farming in the US was the geographic concentration of the broiler industry. This concentration commenced in the 1920s, and by 1990 Arkansas, Mississippi, Georgia and other South Central states accounted for nearly one-half of total US commercial production, followed by the South Atlantic – and Delmarva – at two-fifths of the total. The reasons, according to the US International Trade Commission, were "relatively low land and labor costs, ample feed supplies, proximity to major metropolitan consumption centers, and the historical development in each region of a vertically integrated broiler production and support network."[45] Virtually all broiler production and more than 90 percent of turkey production was to be found in factory farms by 1990.[46] US poultry production exceeded $13 billion in that year, with exports of poultry a record $774 million, or about 6 percent of production. At the same time, per capita poultry consumption reached a record 41.1 kilograms. The US claimed the largest and most advanced industry in the world, accounting for approximately 30 percent of total world production, more than three times the share of the next leading country (the Soviet Union).

Industrial Selection

As H. W. Cheng reminds us, chickens, as well as other animals, have the ability to change their behavior (behavioral plasticity) and physiology (physiological plasticity) based on costs and benefits, to fit their environment (adaptation). Through natural selection, the population preserves and accumulates traits that are beneficial and rejects those that are detrimental in their prevailing environments. Natural selection is slow but constant; working over multiple generations, the changes to the population often appear silent or undetectable at a given point in history.[47] With change imposed upon them, the broilers have experienced, instead, industrial selection.

The principle of domestication of chickens and other farm animals is similar to that of natural selection: selecting the best animals with the highest survivability and reproducibility (artificial selection). But the process of artificial selection is motivated by human needs and wants – profit, hunger, profit, a desire to prove anything is possible, and profit again. In the case of meat animals, the goal of domestication and breeding is to produce creatures that fatten as quickly as possible, with fewer inputs, and accordingly with less attention to aspects of animal welfare. In the last century, the rapid results with poultry are clear. The development of breeding programs and the emergence of specialized breeding companies founded on genetics has pushed the process faster. If jungle fowl lay 4 to 6 eggs in a year, an industrial chicken produces more than 300 eggs a year. Cheng suggested further "improvements" are possible in breeding for speed and efficiency of meat production. Yet he called for the need to consider the prospect of optimal welfare, with attention paid to resistance to stress, disease and other negative factors.[48] Largely, few breeding programs have done so, although in the European Union some effort has been made in breeding facilities and in statutes to require that breeding consider animal welfare.

If most of the literature on European construction and enlargement after the fall of the Berlin Wall has focused on political and economic integration, there has also been some attention in the literature to the role of science and technology in these processes. This work focuses on environment, space, computers and artificial intelligence, agriculture and other fields. In agriculture, the development of the EU has gone hand in hand with the creation of institutions, protocols and other regimes to ensure food, consumer and animal safety. In the EU, a strong belief still operates in the food industry that democratic institutions with public access to the policy process will lead to better science – and better quality of life; in the food sector, this should mean more humanely produced meat – from the points of view of the bird, the farmer and the consumer.

In the broiler world, since 2000, the EU nations have set out to supervise and regulate breeding, stacking and housing, protection, safety, feeding and additives, and other aspects of production, toward the goal of safer, more healthful foods. Since 2007, as published in the *Official Journal*, the breeding of broilers has fallen under a European directive of animal protection that entered into force in 2010. Yet, for all of its good intentions for the bird, consumer and producer, the directive ignored a series of key official scientific recommendations, by allowing, for example, the stacking of chickens at nearly 22 per square meter – overcrowded – through two derogations, instead of the target of 17 chickens/m^2. It does not require access to an enriched environment, or limit the rate of growth of birds. France itself, now the third-largest chicken producer in Europe, raises more than 80 percent of its *poulet* at the maximum allowed by European regulations.

How can we, indeed, consider the chicken's welfare in light of the industrialization of its production and the pressures on birds to abandon what had been considered their usual habitat and normal behavior? The neuroscientist Lori Marino has provided a complete foundation on which to answer this question and to conclude that broiler chickens are chickens, but are *manhandled* in the industrial world. Marino observes that domestication, and recent "very intense breeding and genetic manipulation directed toward production traits such as egg laying and growth," have not been accompanied by evidence of changes in the cognitive or perceptual abilities of domestic chickens (nor has it happened for most other animals domesticated for food). She writes, "Social groups of jungle fowl and wild or free-ranging domestic chickens usually consist of one dominant male and one dominant female. Within their home range, they have regular roosting sites, including high up in the branches of trees." Diet is highly varied and ranges from berries and seeds to insects and small vertebrates. In batteries and cages, their diet is scientifically adjudicated to ensure rapid fattening, and there is no hunting, pecking or moving for seeds and insects.[49]

Marino conducted a meta-analysis of chickens that indicated a wide range of cognitive, emotional and behavioral traits and abilities in chickens that most people do not recognize or acknowledge. For these people, roosting and pecking are usually a cause of amusement, or they see the remnants of fowl – defeathered, deboned, deskinned, in parts and pieces, wrapped in cellophane packages with nary a speck of blood – no longer as animals, but like automobile parts. As the most abundant of all domesticated animals with almost 20 billion worldwide alive at one time, she suggests that there should be greater recognition of the birds' intelligence. As Marino points out, since chickens are categorized as a commodity, they are "devoid of authenticity as a real animal with an evolutionary history and phylogenetic context." Indeed, many people have little belief in the chicken's sentience or "mind." The chicken is "misperceived as lacking most of the psychological characteristics we recognize in other intelligent animals and are typically thought of as possessing a low level of intelligence compared with other animals." The source of much of this thinking may be that "the scientific literature on chicken cognition and behavior is relatively sparse in many areas, and dominated by applied themes, artificial settings, and methodologies relating to their 'management' as a food source." Their welfare is less important than their productivity. Finally, most studies on animal intelligence focus on primates, dolphins, elephants, and only such birds as crows and parrots. Yet it makes sense that chickens should have much in common with other birds and, generally, as animals, they should have complex intelligence.[50]

Chickens are sensitive and curious animals with well-developed senses of touch, sight, smell and taste. The beak is a complex sensory organ with numerous nerve endings. They use it to eat and grab non-food objects to nest and explore. Debeaking thus is painful, and debeaked chickens are defensive, and less able to peck and preen. Chickens have well-developed visual abilities for cognition and spatial orientation. They can make choices when vision is partially occluded, and this is why in industrial farming they are kept in large, dark sheds to keep them as immobile and uncurious as possible. They can detect sounds in a variety of frequencies. They have ordinal abilities. All of these are important to the chicken's social behaviors that the poultry managers and scientists have studied in order to limit the focus of the chicken's brief life to meat production and egg production. Make no doubt about it: chickens are animals, with a full sense of intelligence, learning, memory, sociality and other characteristics. Some studies reveal chickens also demonstrate self-control, self-awareness, reasoning and logical inference, and social learning, for example, as revealed in their setting up of a dominance hierarchy – a pecking order.[51]

Chickens can see a broad panorama, and they have an acute sense of color. They select food based on visual and immediate taste cues. If the composition of the food changes due to availability of grains, the hens may have difficulty shifting to eating seeds of a different shape or color. Their hearing also is acute, and they communicate within and among flocks of birds in a variety of ways. They use postures and displays to signal threat and submission. They have a variety of calls of warning and predator alarm, contact, territorial, laying and nesting, mating, threats, distress, alarm or fear, contentment and calls for food. The new means of raising chickens for meat interfere with these well-developed senses, ways of communication and social structures that chickens develop.[52]

Chickens communicate in a sophisticated fashion with a series of "distinct vocalizations" and "visual displays" that convey information, and not only simple responses to stimuli. Negative and positive emotions range from fear "including capture and restraint, open fields, and novelty" accompanied by physiological reaction, to contentment accompanied by "comfort behaviors (e.g., preening, wing flapping, feather ruffling, body scratching) consistent with relaxation," to empathy. On the basis of this research, it is safe to argue that chickens possess a variety of capacities that indicate they are highly intelligent. One need not push the extensive evidence to point out that chickens are like other birds in "their ethological complexity."[53]

Chicken Ecology and Welfare

A significant and growing literature considers the welfare of animals under factory farm conditions. This literature engages implicitly and explicitly the preceding historical review of literature concerning chicken ecology, habitat and behavior. Specialists in ethology, philosophy and other fields have contributed to the rise of this science. The rich discussions in *Journal of Animal Welfare*, published since 1992, the *Journal of Applied Animal Welfare Science* from 1998, and, from 2014, *Frontiers in Veterinary Science – Animal Behavior and Welfare*, among other forums, indicate the vibrant development and future of this field. I take it as a given that there is a field of animal welfare research, although its specialists inform us of the difficulties of research here – for example, in positing that animals feel and suggesting how to measure these feelings, and even the difficulty in studying animal consciousness. My conclusions here are not about that field and its important contributions. Rather, I review this literature to suggest that animal welfare – an implicit recognition of animal consciousness – must be a part of animal ecology, and that, to study the environmental consequences of factory farms, we must consider the conclusions of these researchers in order better to understand the implications of the physical circumstances in which birds are raised and harvested, and of the genetic manipulation of birds.

Researchers have observed a series of abnormal behaviors among caged birds.[54] In high-density cages, the birds and feathers make up a higher proportion of stimuli relative to the litter area that otherwise draws the natural pecking. It is possible that the birds may perceive the feathers as dust, and this can cause a redirection of ground-pecking to feather-pecking.[55] A similar problem is cannibalism. If some research is directed into alternative husbandry systems to replace the cage, then very few birds are raised in free-range conditions, as this is uneconomical in terms of labor, food requirements, presence of predators and disease control. Yet naturally it is clear, and in environmental preference studies it has been proven, that hens, given a choice of a cage or an outside run, choose the run. Hens used to living outside in the garden all chose the run, while hens previously used to living in cages tended to choose the cage on first trial, although subsequently they came to choose the run.[56]

Since we cannot be certain about whether animal consciousness exists or what it is – without fluently mooing, growling and clucking with them – then how do we approach the problem of defining animal welfare and understanding how CAFOs might indeed be a violation of their welfare? Speaking about meat animals, Dawkins and Layton suggest that the path to take is to give up on moral arguments that animals matter because they consciously suffer, and to appeal to self-interest. They say that treating animals badly in terms of their welfare is bad for our welfare in a variety of measurable ways: in terms of disease transmission and nutritional deficiencies. If we care about our health and well-being, then we ought to care about animal welfare. Also, the behavior of a vast number of devoted pet owners points to the fact that many of us believe to some degree in animal consciousness, and that skepticism about its existence cannot therefore be why we do not treat (other) animals better. Finally, they insist, our treatment of animals is embedded in the problems of global warming and food production. Thus, ecologically, our welfare is connected to the welfare of animals, and we must think about the welfare of everything to which we are connected to save ourselves ecologically.[57]

The pushing of animals harder and harder to serve their function for people, as meat animals that we assume are suffering, has stimulated the rise of animal welfare doctrines that have a place in factory farms of all sorts, whether for cattle, pigs, chickens or other animals. In a study of beef CAFOs with relevance to broilers, Timothy Pachirat worries what occurs to humans as well as to meat animals in modern-day factory farms. He notes that the morals of factory farms are to serve industrial efficiency. Pushing animals into tightly controlled spaces, and engineering their bodies to become meat machines, we necessarily deprive them of their welfare, no matter what the regulations to avoid these circumstances are. CAFOs, thus, are a place of brutal, efficient, out-of-sight, and thus out-of-mind, dispatching, except among those poorly paid workers responsible for killing, skinning and dismembering – these are the words of Pachirat as he describes their "monotonous mechanistic murder,"[58] without concern for animal welfare or our own morality.

Take but one approach to the animal welfare problem. Dawkins and Layton argue, not surprisingly, that progress in improvement in animal welfare "is hampered currently by the seemingly conflicting demands of welfare, commercial production, food security and calls for increasing intensification to curb climate change." If a major goal is for animal welfare to be improved on a commercial scale by future breeding programs, then the chicken industry must seek to breed broilers with high welfare traits, that do not require feed restrictions, that produce birds with low disease levels without the need for routine medication, whose chicken meat is healthy and good for humans to eat, produced in systems that are environmentally sustainable and in an economically profitable way. But achieving that kind of bird unit within a production ethos is nearly impossible. First, a high juvenile growth rate to fatten broilers within seven weeks is incompatible with good welfare. Second, feed restriction in adults is inevitable with fast-growing juveniles. Dawkins and Layton believe that there is a way out, a technological solution, that the conflict between good welfare and productivity can be tempered by making use of all available genetic variation from existing breeds and other sources, and then selecting birds for the range of environments they encounter in commercial (CAFO) production.[59] But their position, I believe, assumes a technological solution to a problem of technological origin within a system founded on a productivist ethos, and hence likely an impossibility.

The incremental effort to scale back technologically driven mandates in bird raising and management reveals the challenges facing factory farm managers, regulators and consumers in achieving the goal of producing meat in better conditions for birds. New, "compassionate" welfare standards in the EU, adopted in 2012, magnanimously added "an extra postcard-sized bit of shared space" to the A4 paper size they already occupied in the barren battery cages of egg-laying hens. Yet half of all eggs in the EU are still produced in cage systems. If full debeaking is no longer allowed, then beak clipping is still permitted – it happens at a day old, and it is likely little consolation to Chicken Little that farmers must use infrared lasers rather than the hot, cauterizing blade of before. The birds produce 320 eggs in 72 weeks. This intensity of production affects the bones that become brittle and break easily, the birds become stressed, and so they lose their beaks. The US standards are worse, and so the EU will not give US eggs or meat visas to enter the Union.[60]

Swedish animal welfare statutes offer, at first glance, a counterintuitive solution to concerns about animal welfare. They are based on the notion that one can limit the spread of disease from animal production with *less* intervention, "by adopting animal welfare legislation that seeks . . . to ease the industrial pressure on animals." This means rejecting not only cost considerations as the major concern for those raising animals; not to feed anything and everything to animals to lower feed costs; not to pack animals tightly into more dark and dank spaces; and not to use antimicrobials, if possible. The Swedish case is instructive here because many people consider Sweden a world leader in animal welfare legislation, including the 1988 Swedish Animal Welfare Act whose regulations focus on prevention of cruelty, provision of adequate food, water and shelter, and address issues of housing, restraint, health care and transportation. The goal is to raise animals in as "natural" an environment as possible, and to allow for natural species-specific behavior: access to pastures for some animals, perches for chickens; prohibitions against such "cruel" tools as driving instruments and tethers, and so on. Other regulations require that sick and injured animals be treated quickly; that drivers receive special education; and constant effort to prevent disease, injury and discomfort. Already, this is usually easier to achieve in Sweden because of the size of operations; in Sweden, the typical swine herd is around 300 pigs.[61] The larger the industrial flock or herd, the more difficult it is to manage with animal welfare in mind. The US produces 115 million pigs annually, at an average of 2,000 pigs per farm, the largest number of them raised in CAFOs.

If Swedish standards suggest a workable alternative to industrial production of meat animals, then within the European Union those standards are not as strict. These, in turn, are much more rigorous than those in the US, which appear in comparison to be almost non-existent. The EU standards specify minimum space requirements for chicken CAFOs – although that space is vanishingly small. For example, recently – in 2016 – in Côtes d'Armor, France, about 50 animal welfare checks were carried out by the local departmental inspectorate at farms with more than 500 broilers in confined systems. In some places, inspectors noted excessive lighting. In other facilities, the birds were too tightly packed, although inspectors concluded that at no time does the loading exceed 42 kg/m². In the majority of cases, violation of the rules, the inspectors claimed, was minor "with overruns of less than 5% density or one batch out of seven with a mortality rate that is too high." (One of seven – 14% – is a huge number of birds.) New French procedures for broiler farming with a density greater than 39 kg/m² are supposed to rectify the situation, and any excess of the mortality rate must be explained, in the face of potential sanctions. It may help that facilities receive notification of the inspection date beforehand so paperwork can be prepared. Yet it seems that the French – and other EU members – are slowly making progress to limit the concentration of animals per square meter.[62]

Of course, appropriate animal welfare involves housing, feeding, disease treatment, stocking density and, increasingly, also genetics. The nature of the health, wholesomeness and safety problems facing the poultry industry has changed significantly in the post-war world with the expansion of cage, battery and shed feeding, tighter packing, and new disease vectors, to include genetic selection of poultry. In one study, researchers asked whether it was assuredly safer and better to pursue industrial production as opposed to the barnyard raising of birds. Do the genetics of today's fowl, which is focused on rapid fattening, drive out all other considerations about animal welfare? If the goal is low-cost and rapid fattening, and if the new broiler models permit this approach, then why would producers seek more expensive free-range, organic, or any other more costly approach? Or has the industrial imperative and concentration of birds among food, droppings, and feathers led to such new and entirely unresolvable welfare, safety, health and environmental problems that another approach must be taken?

A root of this issue is tied to a technological imperative. Animal welfare commences earlier than at the factory farm: it commences with genetics and breeding in the laboratory. Modern breeds of broilers have been heavily selected for high juvenile growth rate, breast meat yield and efficiency of feed conversion. Yet this has left them susceptible to cardiovascular disease, lameness or difficulty in walking. The manufacture of birds into meat machines has increased their physical fragility in many ways. The rupture of the gastrocnemius (calf) tendon in chickens has been a known phenomenon, not a new condition, for some time. But it increased in frequency in the 1950s and early 1960s, almost certainly due to the pushing of birds into technological postures of fattening. The condition results in lameness, and chickens, unable to hold full weight, rest on their hocks; this can lead to swelling and atrophy. Some observers studied whether the condition arose from wire floored coops and those with deep litter, or seasonal factors and nutrition. But veterinary researchers concluded that "it appears to be more the result of genetic selection for fast growth, increased weight, and broad-breastedness, coinciding with accidental stress factors."[63]

An entire productivist ethos undergirds genetics and breeding, which essentially precludes welfare in practice, even if producers are well-intentioned toward welfare. According to Hiemstra and Napel, breeding companies consider welfare not of the animal per se, but as trouble-free production, the absence of abnormalities that hamper production, low mortality and good performance. They note ample evidence that a genetic predisposition to specific welfare problems may be masked by the "favorable conditions in the higher levels of the breeding pyramid, but expressed in commercial production systems that are less well-controlled." The major companies – Aviagen (Arbor Acres), Cobb-Vantress, Hyline and others – select for abundance of meat, if also trying to consider leg strength, heart and lung fitness and other characteristics. But in balancing reproduction traits, welfare traits and broiler production traits, it would appear that changes in breeding goals in favor of welfare traits will only occur within market requirements[64] – "the market" demands rapid growth, and consumers demand meat.

The breeding hens – not only their offspring – suffer from significant welfare problems as well, even if they live longer. Dawkins and Layton write how breeders rapidly become obese; the males have reduced fertility and are reluctant, or find it difficult, to mate. The females have multiple ovulations and decreased egg output. The manufacturers strive to avoid these negative symptoms in practice by restricting the amount of food that the growing breeders receive by 25–50 percent of what the birds would normally consume if fed on demand. If restricting the amount of food available greatly improves the physical health and reproductive potential of the birds, then it also raises the welfare problem of the birds exhibiting signs of chronic hunger.[65]

Eventually, EU officials and parliamentarians called for a more concerted, consistent and direct effort to counteract the animal welfare risks associated with the genetics of factory-farm production. The European Food Safety Authority (EFSA) determined that the genetic improvement of broiler performance had been accompanied, not surprisingly, by an increase in welfare problems, as the animals have been pushed harder to fatten more rapidly. It advised breeding companies to apply a "balanced breeding program" and recommended that birds requiring fewer feed restrictions should be used as future breeders.[66] In fact, all food production in the EU is subject to the General Food Law (EU 882/04) that requires attention to be paid to safety management and the other legal issues, namely: protection of consumers' interests, fair practices, the protection of animal welfare including health, plant health and the environment. Indeed, Council Directive 98/58/EC states that "no animal shall be kept for farming purposes unless it can be reasonably expected, on the basis of its genotype or phenotype, that it can be kept without detrimental effect on its health or welfare."[67]

Making these concerns more complex is the fact that "a substantial proportion of all poultry breeds is endangered, with commercial poultry pedigree stocks in the hands of a limited number of breeding companies"; rural poultry are kept by a variety of breeders, and hence less in danger of becoming complete monocultures. Three companies dominate worldwide the market for poultry meat breeding-stock: Aviagen Broiler Breeders, Cobb-Vantress and Hubbard. The EU Parliament intends, through legislation, to promote free trade in "breeding animals" to ensure the sustainability of breeding programs and preservation of genetic resources, although fowl are not covered by EU legislation, or, apparently, in national legislation.[68]

Can There Be Modern Chicken Ecology?

There are other ways in which the factory-farm system creates tension with health and environment that also have technological roots. They concern not so much raising birds in factory settings, or, before that, ignorance of chicken behavior, but the processing, transport and sale of broilers. We find strong evidence of this problem in the history of the differences in safety standards and quality of food in the US, Europe and elsewhere. A constant push in the US to find a technological solution to the dangers of disease, spoilage and other aspects of chicken meat production endemic to factory-farm production prevails in the US, whereas in Europe the attempt is to use more commonsense and low-tech approaches to securing public health benefits. It must also be kept in mind that safety, wholesomeness and public health are not simple determinations based on some universal scientific language, but reflect national and local determinants, values and considerations, and, as a result, outbreaks of such diseases as Avian flu, or scandals involving unadulterated products sold as safe, will more likely occur, and in some settings are more likely than in others – even if the broilers involved are genetically the same.

An example of this concerns egg production. While not a focus of this book, it is instructive briefly to consider the difference between EU and US rules on eggs. The approach is so different that European supermarkets generally do not refrigerate eggs, and need not. But the USDA requires that supermarket eggs – graded eggs – be washed and sanitized before sale to reduce the risk of *Salmonella*. European regulators note that washing may in fact contribute to the transfer of harmful bacteria from the outside to the inside of the egg. Yes, the

Salmonella bacteria come from two main sources, from the laying hen herself if she is infected, or from the hens' feces. The US adopted its regulations because eggs are produced mainly in enclosed, large-scale houses where disease, indeed, may rapidly spread. Hence, they are required to be washed along conveyors. But the washing enables bacteria on the wet shell to grow and penetrate. In Europe, regulations are intended to produce clean eggs at the point of collection, not later, and this encourages "good farm husbandry and practices." Even more, washing may damage the outside layer of the shell, the cuticle, which may permit bacteria, again, to penetrate. Against this eventuality, the US require that washed eggs be refrigerated, and, once refrigerated, they must remain refrigerated. On top of this, many egg farmers in Europe vaccinate their hens. The outcome is that *Salmonella* is the most common cause of food poisoning in the US, according to the US Food and Drug Administration (FDA). According to estimates, more than 140,000 people get sick each year from eating eggs contaminated with the bacteria, which triggers non-life-threatening (though unpleasant) symptoms like diarrhea, cramps and vomiting.[69] It does not help that, in the US, among the major scofflaws of standards and cleanliness in food industries have been egg producers.[70]

In light of the tens of billions of fowls raised annually throughout the globe, most intended for faraway markets and distant consumption, how is it possible to ensure food safety and animal welfare? The sheer number of birds – or other meat animals – raised in conditions ripe for the spread of diseased flesh and microbial infections – makes the task of ensuring consumer safety challenging, no matter where inspections must occur. But it seems that a model based on common sense and moderated industrial approaches that give animals at least some possibility of being raised in "natural" conditions will be preferable to brute-force efforts to kill the foodborne diseases that arise in factory farms – for example, harsh chemicals (chlorine) and antibiotics.

The alternative to farm factories and CAFOs, as a way to raise meat animals but to allow their behavioral traits to be manifest as they grow toward their final fate of being food, is found in what is called free-range raising of fowl. The Cobb Creek Farm in Hillsboro, Texas, for example, celebrates pastured poultry that is "raised the old fashioned way: on fresh green pasture and wholesome, non-GMO feed. Our model has been developed over the last twenty years and allows the birds to be raised in a cleaner, healthier, happier environment. All Cobb Creek boilers are harvested humanely right on the farm in our state-inspected processing facility." The chickens are raised humanely, and given "the continuous opportunity to act on . . . natural chicken instincts." Brooder houses are clean, quiet, aerated; spacious mobile range coops enabled "fresh air, forage, exercise, sunshine, and a non-GMO grain ration."[71] And slaughter is one-on-one, not mass production – although I am not sure that the chickens know the difference.

3

Chicken as Machine

> It is not enough for a sage to study Nature and Truth; he must dare to say it in favor of the few who want and can think; for the others, who are voluntarily Slaves of Prejudices, it is no more possible for them to reach Truth than for Frogs to steal.
>
> – La Mettrie, *Man, a Machine*[1]

In 1747, the French philosopher and physician Julien Offray de La Mettrie, published *Man, A Machine*, in which he asserted that animals were automata, soulless, mere mechanical vessels that responded to external stimuli, and did not feel pain when cut, hit, beaten or butchered. The sounds they made were only simple machine-like responses to physical pressures, a kind of Newtonian "equal and opposite reaction" to a bite, kick, swinging club or prodding knife. La Mettrie's sense of animals has come to fruition – with the exception of the fact that animals have sensory feelings including pain, and emotions as well, and they suffer the consequences of being animals, not machines, in modern agriculture. Through a series of breeding, engineering, filing, clipping and other actions, chickens have become meat machines, geared only to chicken up as quickly as possible, and denied other animal-like behaviors such as pecking with a beak, all in the interests of saving energy for fattening and avoiding damage to the meat vessel.

6 *Machine-like chickens as part of the industrial machine, hooked to an overhead conveyor. Production of white meat.*

Domestication of animals has a long history. Farmers, pastoralists, agriculturalists, collectors and others have long sought to settle animals for purposes of food, entertainment and comfort. Selective breeding permitted enhancement of those qualities desired for better taste and more meat, increased amounts of and fattier milk, more pronounced and vibrant colors, calmer and more "human" behaviors, and so on. But the impetus to industrialize birds, to make them components of a long supply line in a mass-production facility – a large-scale chicken coop, if you will – dates to the time of Henry Ford and his conveyor system for production of Model T and Model A cars. Although chicken was still not an exceedingly popular meat until after World War II, the creation of chicken as a machine commenced in earnest when calculations of optimum temperatures, minimum space and chemical make-up for meat birds were tied to optimum bird parameters in order to hatch, raise quickly, slaughter even more quickly, drain, eviscerate, slice, dice, fabricate, pack and transport birds far and wide to a variety of different audiences and tastes – by the end of the century, across the globe. Establishing the parameters for efficient industrial production required the input of research laboratories at the state and industrial level to retool the chicken itself, rebuilding it from the ground to the beak, but especially subordinating its breasts and thighs to the task of fattening as quickly as possible.

While broiler chickens at one time took 4 months – 16 weeks – to reach a sale weight of 1.15 kg, today they reach their weight of 2.3 kg in half the time – and faster. While the chickens ran around a simple farmyard and were forced into a coop at night lest the fox nab them, today perhaps 50 billion broilers are harvested annually and live essentially their entire lives in temperature-controlled sheds that hold tens of thousands of fowl. Today's broilers grow much larger than the birds of yesteryear due to selective breeding, feeding and environmental control. Alabama-based Aviagen, for example, developed the Ross 308 broiler, a breed that grows to be about four times the size of a comparable 1957 breed and has three times the "breast conversion rate," a measure of grams of feed to grams of meat produced. In about 5 weeks, after eating 7.77 pounds of feed, the Ross 308 grows to roughly 5 lbs. According to the journal *Poultry Science*, if humans grew at the same rate as these modern chickens, a baby would weigh 660 pounds (300 kilograms!) at 2 months.

Of course, chickens are not machines, and they suffer the consequences of being treated as such. Intensive chicken farming that uses three kinds of husbandry in many cases fails to account for their senses of sight and hearing, because the habitats used are overpacked, noisy, and have artificial lighting: cages, meat chicken sheds and breeder sheds. In cages, chickens are kept in groups of three to ten birds, with space allowances ranging from a measly 350 cm^2 per bird in the United States, to as high as 700–800 cm^2 in Norway and Switzerland. In meat chicken sheds, tens of thousands of broilers housed on litter grow in either semi-enclosed or environmentally closed houses. And in breeder sheds, flocks of several thousand chickens live in semi-enclosed or enclosed housing on litter or wire, with a space allowance of 0.2–0.3 m^2 per bird.[2]

Many other creatures became industrial food commodities for similar purposes. Chinese pig breeds came to Europe and later America. Europeans introduced the domesticated Asian hogs that had been bred to feed on waste and agricultural byproducts, to replace wild European pigs as European agriculture intensified. The Chinese pigs were "imported to create improved varieties first in England and then in America. These new breeds, with their enhanced capacity for rapid weight gain, played a vital role in the pig's transformation from a small-farm subsistence animal into an industrial meat producer." The pigs served as a "microcosm of early modern globalization and the emergence of industrial capitalism." With the industrial revolution in Britain and then America, the newly introduced Chinese pig did not serve as a small farm animal, but was transformed into an industrial commodity.[3] Chickens followed a somewhat similar path, although their potential as meat machines was recognized later, and the eating habits and behavioral traits of fowl likely contributed to the later recognition.

The mechanization of poultry farming reflects a number of forces from the business, entrepreneurial, engineering, labor and cultural worlds. It reflects the innovativeness of humans who wish to replace very hard labor with a wide assortment of machines, engines, devices, electrical motors, blowers, new materials and, ultimately, industrial plants and animals that can be more easily planted or inseminated, as seeds or eggs. These animals and plants, raised and fed, pushed and prodded, given antibiotics or weeded, watered and given light, but at data-driven times, are carefully monitored and controlled as if in a panopticon of agricultural production. They are then harvested and processed, the "chaff" and viscera separated from the good grain or meat, weighed, wrapped, boxed, loaded and sent to distribution points ("supermarkets," themselves the apogee of industrial food consumerism). As one celebration of the confluence of the living and the machine noted in 1952, "Chicken chores in olden days usually were women's or children's work, but now these are a major expense problem to large-scale poultrymen. Many of them hope that hens can be made to live an almost completely machine-kept existence, and some of the larger chicken farms have almost reached that happy goal."[4]

In their natural environments, chickens forage, peck, scratch, preen and dust-bathe. They perch in trees. But, as meat birds, they are born in a hatchery, moved to growing buildings, and are regulated in movement, food, water, light and temperature until they are taken out for slaughter. Over 70 percent of broilers in the world are raised in factory farms where they live in sheds. The floors are covered in litter to absorb shit and piss; the litter is removed only after the birds are removed. The broiler is itself a machine. It is incubated, and vaccinated against *Salmonella*, Newcastle disease virus, infectious bronchitis virus, avian pneumovirus, infectious bursal disease and Marek's disease via sprays into its drinking water. It is transported in boxes to the rearing farm, placed on a conveyor, and housed in factory-farm buildings. In the US, there are no limits on stocking density, but in the EU, and in the UK – in early autumn 2019 still a part of the EU, but with different approaches to stocking issues – this is somewhat restricted.

Before transport to slaughterhouses, broilers are deprived of food for several hours. This to reduce the risk of fecal contamination during processing and evisceration – a kind of chicken colonoscopy, and what industrial manufacturers of birds call "gastrointestinal cleanup." The broilers are swept up and caught by machines or workers, pushed or shot into boxes, loaded onto trucks, and sent off. On arrival at the abattoir, they are stunned before slaughter by electrical stunning (an electrified water bath) or gas stunning (CO_2 or inert gases) to render them unconscious, or experience controlled-atmosphere killing like a gas chamber. In the US, low atmospheric pressure systems are now used that mimic slow decompression with removal of oxygen, with resulting loss of consciousness.[5]

Writer Paul Crenshaw describes the clinical horror of chicken abattoirs. The chickens are attached to conveyors upside down, "stunned," cut, drained, scalded, then vented and drawn, their guts opened to remove the edible offal – heart, liver and gizzards – while the head and neck are cut off. Crenshaw observes:

> The birds are then washed, chilled, drained, and frozen. So, to review: hung upside down, shocked and stunned, throat cut, scalded, defeathered, vented, drawn, decapitated, washed, and prepared to be shipped to all parts of the country where someone like you or me is browsing through the meats section at our friendly neighborhood supermarket, wondering what to feed our children for supper.[6]

In any event, the technochicken is here. It is an industrial bird, tied to Fordist attitudes about assembly-line mass production and to factory structures in life and death. The broiler is comprised of feathers and feed, biology and engineering, breasts and frame, bloody in its life and slaughter, bloodless in its packaging, manipulated and accelerated to the summit of biological productivity via the systematic application of science and technology.[7] Treated as a machine made up of various technological inputs, pushed along by technological imperatives, the factory-farm broiler has presented the nations of the world with cheap protein, but also with food-safety and disease risks. La Mettrie's chicken lives with us.

The Chicken Assembly Line

To raise birds quickly and efficiently, specialists have strived to control all aspects of chicken-ness – from the bird itself to facilities where they are raised, from feed to lighting – and, in so doing, have had to ignore those aspects of chicken ecology that were well known and celebrated by researchers from the early 1900s. Here, we consider the "inputs" into the broiler machine, from food to light and ventilation, from sheds to conveyor belts.

Specialists addressed feed early on. On the eve of chicken industrialization, avian researchers determined that soy, corn, other grains, grasses, plus milk products, beef scraps and the like were the key to healthy and meaty birds. They worried that "practical poultrymen" were focusing too much on the protein content, while study of mineral matter and fat had lagged.[8] But in the post-war years, researchers and companies learned that such feed crops – soy, corn and grain – could be provided and manipulated in scientific fashion to speed up fattening. Many of the feeds in use today are GMOs.

The result of abandonment of free-range feeding for scientifically balanced feed has been that the size of broilers has increased, the amount of feed required to produce them has reduced, and the time of fattening has significantly shortened. In 1955, when modern integrated broiler complexes were being established, it took 73 days to produce the average broiler, which weighed 3.1 pounds, and every 100 pounds of broiler production required 285 pounds of feed and 4 hours of labor. By 1980, it took 52 days to produce a broiler that weighed about 4 pounds, and every 100 pounds of broiler required 208 pounds of feed and 30 minutes of labor. By 2006, it took 49 days to raise the average bird, which weighed 5.4 pounds, and producers used 195 pounds of feed for every 100 pounds of broiler production.[9]

At one time, industry representatives considered the use of hormones to accelerate growth. But the use of hormones for poultry in the US has been banned since the 1950s. Aside from being illegal, they would be difficult to administer, would be broken down through digestion if fed to broilers, and injecting tens of thousands of them daily would be impossible even for a drug addict with experience. The US FDA approved the use of supplemental hormones in beef production as scientifically proven to be safe. Beef industry people note that hormones exist naturally in plants and animals. The six hormones approved for use in beef production include both natural and chemically similar synthetic hormones that are primarily administered using a pellet implanted in the back of the ear, with hormones released over time to ensure low concentrations, and with the ears cut off and discarded at slaughter so no one eats the pellet. The hormones increase the speed and volume of meat production, thereby lowering production costs. People who eat more beef are at higher risk from cancer.[10] But the ear of a chicken – which indeed exists – is not large enough to implant a hormone pellet in, and there are a number of other ways – with feeds and antibiotics – to stimulate rapid growth.

Chickens themselves are not yet genetically modified, even if their feed is, and the industry believes therefore they must not – and need not – have any special labeling as genetically modified organisms (GMOs), even when they consume GMOs.[11] According to the Non-GMO Project, it would be better not to permit the pervasive use of GMO feeds throughout agriculture. The Non-GMO Project notes that, in the US, "60% of all cropland is used for three crops: corn, soy, and forage. More than half of that corn and soy is used for animal feed, and almost all of it is GMO. . . . With the introduction of genetically engineered alfalfa in 2010, forage acres are also increasingly at risk for being GMO." The Project opposes these GMOs for a variety of reasons, whether or not there are direct human health impacts. Its specialists write:

> Almost all GMOs are engineered for one of two traits, or a combination of both. These traits are 1) herbicide tolerance, where a plant can be sprayed directly with a chemical pesticide like Roundup and survive, and 2) insecticide production, where a plant produces the Bt toxin so that if an insect eats any part of the plant it will die. From a sustainability perspective, these traits may have serious consequences for the environment and for our food system.[12]

Several major manufacturers, Dannon among them, have agreed to give up GMO-fed animals. At the very least, it would be nice to have complete transparency, with labeling on packaging to indicate the various bird inputs, so that consumers themselves can decide what to eat and in what quantities.[13]

The broiler lives, fattens and dies in a carefully controlled built environment. Chickens are sensitive to light in terms of egg production and weight gain. The study of lighting – wavelength, intensity and duration – has enabled some success in determining optimal conditions – for example, by changing photoperiods to influence broiler productivity and health.[14] When the days shorten in the fall and winter, chickens, if they were to live so long, would begin to rest and production of eggs and meat would fall. Thus, producers manipulate light intensity and duration to manage broiler production and well-being. Poultrymen have had experience in the use of artificial lighting since the 1920s or 1930s. We can assume that the bright incandescent lights of that time had a variety of impacts, not all of them optimal for the birds. But the studies of the early years provided a baseline for future production standards. In tests of lighting – from full lighting for 23 hours and 1 hour without, to staggered lighting, but still with at least 16 hours of light daily, with lighting on for some time and then off for some time – they learned that too much light was bad. In a series of studies, researchers sought to identify how much was too much against the final goal of increasing body weight as quickly as possible. But achieving the criterion of "quickly" resulted in such leg problems in chickens as tibial dyschondroplasia, if not directly increasing mortality.[15]

A recent article noted that, despite considerable research into light intensity, there was still a debate on the optimum level to be used for intensively housed broilers. Obviously, insufficient intensity – perhaps employed to keep electricity costs down – was ineffective and dangerous. Illumination at 1 lux in an experiment resulted in heavier wings as a percentage of live weight (1 lux is equal to 1 lumen per square meter.) If light intensity had no effect on skeletal health, then birds exposed to the one lux treatment had heavier and larger eyes and increased ulcerative footpad lesions.[16] What was the proper light intensity? In 2018, researchers determined, through a meta-analysis based on published data, that 5 lux of light intensity during grow-out period should be the minimum level to maintain sufficient productivity and welfare of broiler chickens.[17]

Scientists also claim that broilers see differently from many other animals. Chickens have color vision, and different colors and light levels affect chicken behavior. One Delaware specialist noted that studies using monochromatic light demonstrated superior growth for broilers given blue or green light, with other studies suggesting that broilers are less active under blue or green light than under red or white light, and that yellow–red color may increase activity.[18] For many years it was accepted practice to brood and rear chickens under red light to prevent cannibalism, based on the concept that red light makes it difficult for a potentially cannibalistic bird to see red blood vessels and blood on other birds. Currently, the most useful method for preventing feather-pecking and cannibalism is to house birds under very dim white light. The birds cannot see each other well, and this is thought to reduce aggressive behavior among them.[19] Green and blue light improves growth and lowers age of sexual maturity, while red, orange and yellow light increases age of sexual maturity, and red and orange light increase egg production. Birds are calmer in blue light. In any event, light was an important input because its manipulation might encourage growth, yet tamper with stress and activity, and assist in the management of disease, deformity, and cannibalism due to stress, overcrowding and lack of vitamin D. After all, CAFO chickens may never see the light of day in their sheds.

Nose Job

To live among other broilers, some broilers get nose jobs. In fact, the beak is essentially a touch organ, with extensive nerve endings. It is used for food apprehension, preening, defense, and for the establishment of social hierarchy.[20] Another stage in the rise of the industrial chicken was the development of beak-trimming techniques. As social creatures, chickens do best in groups of fewer than 90 birds, where they can establish a pecking order. In larger groups, they may not succeed at this, and they may as a result peck one another repeatedly to the point of injury or death. To prevent this, but not influence the natural pecking behavior, nor abandon massive and overpacked sheds, farmers introduced beak trimming. Granted, debeaking is normally not carried out in CAFOs since the birds are slaughtered before they are mature enough to begin to engage in pecking cannibalism.

Some of the first efforts at systematic beak trimming were carried out by farmers at the Ohio Agricultural Experiment Station, who hand-trimmed a quarter-inch off the beak in the late 1920s and 1930s. D. C. Kennard, whose research career lasted into the late 1950s, introduced an industrial beak-lopping method, and suggested the use of a sharp knife.[21] As the broiler industry became increasingly industrialized, more machines addressed the challenge of processing thousands of chicks in a short period of time. The "chicken debeaker and inoculator" (US patent #3502079 A) enabled an operator to trim and cauterize chicks' beaks with a heated blade while inoculating them with an atomized spray against the eyes. One operator could perform two jobs at once. James Lyons received a patent in 1971 for this hand-held unit that trimmed and cauterized the beak simultaneously with the application of a dose of vaccine, while the only accessory equipment needed was a low voltage supply for the resistance-heating trimming blade.[22]

Beak-trimming at the contemporary factory farm is performed via more automated methods. The industrial machines – shears – used today are quick and their users claim they do not inflict pain or damage on the chickens.[23] But, then again, researchers have never had their own beaks trimmed. A recent study indicated quite the opposite: debeaking usually involves the beak being seared off with a hot blade precisely to prevent the feather-pecking that can result from the stress of confinement. Debeaking may result in severe, chronic pain, especially since a beak is filled with nerves.[24]

Over the years, in the effort to be more humane, poultry plastic surgeons have developed four other methods: hot blade, cold blade (including scissors), electrical (the Bio-beaker) and infrared. More recently, laser beak-trimming has come into fashion as less painful, providing greater beak uniformity, and better animal welfare. In one study comparing laser versus hot-blade trimming, scientists determined the lasered birds had higher weight gain and better feed conversion ratios. Laser trimming helps avoid poor trimming.[25]

Another method that poultrymen explored to avoid pecking damage was to approximate blurred vision in broilers to prevent birds from seeing each other. In 1962, a farmer in Arizona discovered by accident that a group of chickens suffering from a severe case of cataracts ate less, were easier to handle and got into fewer fights resulting in injury or death to the birds. The conclusion was that partially blinded birds were far more profitable than their fully sighted counterparts. An idea was born – figure out a way to similarly reduce the vision of the nearly half-billion chickens in the US. Optical Distortion, Inc. (ODI) was formed to tackle this opportunity. ODI developed a lens that would partially blind the chickens. Its researchers spent years developing a method for insertion and retention that would make the lenses easy to use. The lenses reduced chickens' field of vision and created distortion; they could see enough to live, but the decreased vision prevented fighting and other unwanted behaviors. Moreover, the lenses were colored

red, which, the inventors believed, calmed the birds. The lenses were more effective than debeaking and required less effort on the part of the farmer.[26] The lenses "reduce food waste, expenditure of energy, noise, injuries, and resultant disease, dust and egg breakage, promote contentment and harmony, and increase the production of eggs."[27]

Industrial Dispatching

As in life, so in meeting its fate, the broiler became part of a larger industrial enterprise that sought to bring speed and efficiency to the process. At factory farms, with the need to kill thousands of birds quickly, the conveyor belt became the central part of the machine. Instead of a man or woman killing the chickens with a knife, today's slaughterhouse workers stand dressed head to toe in bloodstained, but sanitary, outfits, at the kill position with an overhead conveyor. They dispatch the broilers in warm blood with the Chaplinesque pressure of the assembly line that was captured in the film *Modern Times* (1936). Imagine the contrast before the assembly line: a 1927 pamphlet advised on the best way to kill a fowl – specifically, to bleed it by severing the arteries in the neck. First, the plucker suspended the fowl by the feet at about shoulder height. Then, using a "particular kind of killing knife," with the blade heavy steel, about 2 inches long, a quarter of an inch wide, ground to a sharp point with a straight cutting edge, the plucker takes the head

> in the left hand and the knife in the right hand. With the thumb and forefinger of the left hand the mouth is forced open by pressure and the knife is inserted into the mouth with the blade pointing toward the back of the head. The knife is then forced up to the juncture of the head and neck where the arteries come down on each side of the neck; these are severed and the fowl bleeds freely. For a left-handed person the operation would be performed in the opposite way.

The authors insisted that "when this is done properly the bird will squawk and it will also make a convulsive movement which tends to loosen the feathers in the feather muscles. If the brain has not been properly pierced the feathers are hard to pluck and the skin is frequently torn badly."[28] What a joy when the conveyor belt enabled the process of stunning, killing and bleeding with mechanical and electrical devices!

To ensure humane slaughter – yet industrial efficiency without pause – contemporary operations use one of three means to stun the chickens before draining them of blood: electrical stunning, whereby chickens are hung upside-down by metal shackles on their legs and stunned using an electrified water-bath into which they are dipped, before an automated knife cuts the throat and they bleed out; gas stunning, whereby CO_2, inert gases or a combination is used to induce unconsciousness before the killing machine takes over; and controlled-atmospheric killing that actually suffocates the chickens (and avoids the risk of birds regaining consciousness after gas stunning, but before their throats are slashed).

Officials of the EFSA recently commissioned an effort to determine precisely how best to stun broilers before killing them, based on evaluation of "the sensitivity and specificity of indicators of unconsciousness and death," including such methods as penetrative captive bolt for bovine animals and head-only electrical stunning for pigs, electrical water baths, and the now-employed and authorized gas methods for slaughtering poultry. They also considered slaughter without stunning for chickens and turkeys. They evaluated these methods using electroencephalography (EEG) reference tests for unconsciousness and death. Ultimately, not surprisingly since the animals were stunned and spoke another language, the researchers could not draw any firm conclusions.[29]

But in the effort to arrive at some procedure acceptable to the public, EFSA's Panel on Animal Health and Welfare (AHAW) launched an online public consultation for its draft scientific report on the welfare aspects of genetic selection in chickens raised specifically for meat production (broilers), so that laypeople could have input – the idea being that better methods would result from up-to-date and comprehensive scientific advice to EU decision-makers, including that from citizens. Yet, if you do a subject search of the EFSA webpage for health and safety issues pertaining to broilers, you will find mostly articles and studies on animal feed (281), biological hazards (42), GMOs (32) and a few other topics, but very few on subjects directly relevant to animal welfare.

Broilers as Pieces, Parts and Nuggets

We cannot see the disease, nor much of the filth, connected with factory-farm production and chicken, a machine. But the way chicken is handled and packaged both increases the likelihood of the spread of disease at various stages of handling, and it makes it seem as if the foods we buy are absolutely safe. After all, they appear as a variety of different products – parts, portions, sausages and nuggets, in wonderful containers made of modern plastics. The industrial chicken found its way into stores and homes through innovations in packaging and marketing in the 1920s and 1930s, in connection with the expansion of refrigeration, the rise of plastics and the creation of the modern supermarket. The first US supermarkets opened during the 1930s. Consumers flocked to them, although the Depression delayed further expansion until the post-war years, when they became a central feature of life.[30] During the period 1948 to 1963, large chains increased their share of the nation's grocery business from 35 to almost 50 percent. Managers installed high-capacity refrigerated display cases for the myriad frozen and other processed-food products, and increasingly cellophane-wrapped chicken packages.[31]

Plastics are part of broiler history. One of the most important packaging materials for the industrial chicken is cellophane, also used for artificial kidneys (dialysis), which uses the semipermeable membrane to separate cells and molecules by virtue of their different diffusion rates through the membrane. A variety of manufacturers pushed their products that contributed to packaged chicken: supermarket chains; manufacturers of shopping carts (who wanted wide aisles and touted "self-serve"); box, plastic and other packaging manufacturers (who pushed Americans away from bottles, for example, that were sold with a deposit, returned, cleaned and re-used, to plastic that consumers liked for its convenience, light weight and resistance to breaking).

Invested in cellophane from the 1920s, the E. I. DuPont de Nemours and Company pushed and controlled it in a variety of uses. DuPont encouraged both retailers and consumers to see cellophane and self-service as appealing, convenient and profitable. Because of its transparency, heat-sealability, tear strength, moisture resistance, permeability to gases, and resistance to grease and oils, cellophane

7 *Plastic-wrapped cut-up processed chicken choices in a typical supermarket display case.*

revolutionized the food-packaging industry. For shoppers, cellophane advertisements promised greater convenience and cleanliness, without any loss of freedom to see the contents of the package. DuPont managers believed that if consumers preferred wrapped products and served themselves, then more store managers would feel compelled to provide those products. They produced a series of studies and reports to push cellophane. In one, "Self-Service Meats: Progress Report on a Promising New Development," DuPont claimed that cellophane would speed up butchering and that consumers would buy more if they served themselves.[32] Du Pont managed to control 75 percent of sales in 1946, and still 68 percent in 1950, when anti-trust trials were brought against it.[33]

Packaging is an art, an industrial art, a psychoanalytical art, whereby industrial "visualists" manage to probe and unlock our inner desires with suggestions of "bounteous fulfillment" of taste and other sensual pleasures. I am not sure that the cellophane packaging for the seemingly bloodless, lifeless slices of meat and pieces of animals accomplishes this end, especially since the muscle meat, arranged to hide the sinew and fat, offers a grayish or brownish tint. To the many other environmental costs of industrial chicken production must be added the plastic trays and packaging that cannot always be recycled – and few people do in fact recycle such packages of fish, meat or poultry because they find it dirty and, after a short while, vile-smelling, and the convenience of the self-serve supermarket in picking these and other finished goods means also the convenience of throwing away the wrapping without a second thought. Open-style display cases contribute to the sense of convenience, and sale with nary a thought to the energy required to keep broiler products chilled. One historian contends that this kind of modern packaging – whether the clear plastics for meat, chicken, fish, fruits and vegetables, or the colorful boxes of ready-to-pour and mix or consume semi-perishables – contributed to the rapid decline in full-service meat, cheese and produce counters.[34] It may have come with a 15-year delay after the fall of communism in the USSR, or simultaneously with the rise of state capitalism and consumerism in China, but the full-service establishment is slowly giving way to the supermarket in many places in China and elsewhere.

Broilers and Mechanically Separated Poultry

In one last way, we consider the broiler a machine in its presentation as mechanically separated poultry (known by the nutritious "MSP" in the business) products. The broiler, no less than lumber to be sawed and planed, ore to be milled, or Detroit auto parts to be affixed to a chassis, was a tool and the object of tools. The poultry dressing and processing industry, like those for other animal products, became highly mechanized, with labor increasingly efficient at using knives and machines to butcher the animals, and other machines and processes to suck and grind every morsel of flesh, cartilage or ligament – and some bone fragments – for such "value-added" products as nuggets and sausages.

Mechanization of the assembly line – the slaughter, drainage and processing line – enabled value-added products to be a boon to the major producers. In the US, output per employee hour in the poultry dressing and processing industry rose at an average annual rate of 2.9 percent between 1963 and 1985, accelerating in the last years of that period and somewhat higher than the rate for all manufacturing, and reflecting the transformation of processing from predominantly manual to mechanical operations. Automated eviscerating and cutting machines were one key. Consumer demand, higher-valued poultry products, and the fast-food industry pushed increases in output. At the next stage, new processing technologies and streamlined federal inspection procedures also increased output.

Driven by, and pulling, technology, the industry was transformed from supplying mainly whole fresh or frozen birds to the market in the 1950s to supplying parts in the 1960s, and by the 1980s more than two-fifths of the birds were cut into parts. Prepackaged part trays, preformed ground patties, luncheon meats and hot dogs appeared, taking approximately 13 percent of the market by the 1980s. Finally, the introduction of such high-end frozen *entrées* as stuffed chicken breasts, and specialty items for the fast-food and restaurant industries such as nuggets and breast slivers, secured the chicken, as a machine, in the American market.[35]

The demand for fast food or, as manufacturers called them, "convenient and ready-to-eat products," from the 1970s, caused chicken-unit producers to salivate over consumer-oriented value-added chicken products. At first, producers acted on the belief that the consumers' "ideal" chicken product was a refrigerated product made with free-range chicken breast, and produced with no additives, preservatives or flavors. Indeed, surveys indicated that consumers were willing to pay 30 percent more for value-added chicken products over the prices of conventional products, and younger consumers, individuals who shopped at farmers' markets and preferred free-range or organic products, were more likely to pay a premium for value-added chicken products.[36] Like seafood products, so chicken products found a variety of applications and sizes, were manufactured using various extraction, washing, grinding, assembly and cooking processes, with additions of preservatives and flavorings, and served in homes, restaurants and schools.[37] Chicken value-added products are highly manipulated industrial material that is filled with flavors, preservatives, and often with MSP.

One of the most interesting processes is the manufacture of mechanically separated chicken – that is, a product made by separating meat from chicken or turkey bones using screens and filters to remove the bones and bone chips and retain all of the meat. Of course, we have all tasted bone chips in the meat; it's a marvelous tactile discovery. MSP finds a home in hot dogs, lunch meat and other processed treats, but not usually in nuggets or patties, and it is not sold directly to consumers. According to federal law, if MSP is used in any food product, it must be listed among the ingredients as such. According to Meatsafety.org and Poultrysafety.org, websites for the American Meat Institute, MSP is safe and nutritious and may contain slightly higher levels of calcium and phosphorus – both essential nutrients – than poultry meat removed from bones by hand. KFC and Chik-fil-A are two of the ten most popular fast-food chains in the US, and rely on the broiler industry for a variety of their processed meals. See Table 1 for a summary of various industrial chicken products.

Nuggets may be the most familiar chicken fast food to the world's consumers. According to the standard story, Cornell Professor Robert C. Baker, and a student, "mastered two food-engineering challenges: keeping ground meat together without putting a skin around it, and keeping batter attached to the meat despite the shrinkage caused by

8 *The now-ubiquitous nugget that has entered the life of the world's consumers. If done well, and breaded lightly, it is a tasty treat. If not, it is typical factory food.*

Table 1 *Value-Added Chicken Products: The Culture of Fast and Ready-Made*

Value-Added Chicken Products
Re-formed breaded and battered chicken (nuggets, dippers and burgers)
Whole-muscle Nuggets, dippers and goujons
Premium Chicken Sticks (Satay, Yaktori, Tandoori)
Flavored and coated chicken fillets, bites and strips
Kebabs
Skinless and boneless
Thighs, legs, wings
Ground
Marinated
Breast slices
Chicken meatballs
Chicken, cheese and onion sausages
Frozen versions of all the above
Gizzards

freezing and the explosive heat of frying." They ground raw chicken with salt and vinegar to draw out moisture, and then added powdered milk and pulverized grains as a binder. They shaped, froze and coated the sticks in batter and cornflake crumbs, and over time managed to get the sticks to stick together. Baker ran these experiments to recreate markets for white meat that had shrunk with the drop in military demand after the close of World War II. Baker's contribution to bird culture also included chicken hot dogs and chicken steak.[38]

But McDonald's takes credit for the McNugget, and commissioned Tyson Food in 1979 to create the product. The product went on the market in 1980, but was only sold nationally four years later because it took that long to organize supply to meet demand. Suddenly, every franchise wanted them – and wanted more of them. McDonald's marketed them, strangely, as somehow healthier than their burgers, although they were deep fried. The first test marketing of nuggets occurred at a Knoxville, Tennessee, McDonald's in 1980. And according to the National Chicken Council, nuggets are in fact usually made of the same meat that you see in the supermarket – that is, broiler meat, most of which comes from a split breast of chicken or "rib meat," which is the lower end of the breast meat. Other boneless chicken meat, from the legs and thighs, may be added, too, then ground and formed, just like you would form a meatball from ground product. It is then breaded and cooked, usually baked or fried in oil.[39] The most important fact to remember about nuggets is that, during the 2008 Summer Olympics in Beijing, China, the Jamaican sprinter Usain Bolt consumed approximately 1,000 chicken nuggets – roughly 100 and perhaps 4,700 calories' worth of Chicken McNuggets daily.[40]

Of course, nuggets are not all good. An "autopsy" of chicken nuggets was recently published in the *American Journal of Medicine* with the title "Chicken Little." Mississippi researchers, worried about the US epidemic of obesity, and especially at its epicenter in their state, evaluated the nutritional value of nuggets produced in two different fast-food joints in the capital of Jackson. Mississippi, shockingly, prohibits restaurant food restrictions by law. The law does not prohibit buying nuggets, fixing them in formalin, and sectioning and staining them for microscopic analysis. The first sample was 50 percent skeletal muscles, with the remainder primarily fat and some blood, while the second was only 40 percent meat. The processed lean protein had been combined with salt, sugar and fat – actually, over 50 percent fat. Indeed, Chicken Little.[41]

The Broiler in the Aura of Antibiotics and Food Safety

In one other crucial way, the chicken is a machine to be adjusted, pushed and protected by chemical additives – in particular, antibiotics. The treatment of chickens as industrial products means not only that light, heat, feed, metabolism and so on are carefully monitored and regulated, but it has also become standard to treat fowl with antibiotics. Thus, an aspect of the transformation of fowl into units of production was the increased use of antibiotics (antimicrobials) to manage infections in the animals and to hurry weight gain along. Here, medical specialists act as a kind of public health engineer for the broiler. With industrialists having created conditions for disease to become a large problem, with the potential for great losses – illness and death – among family and fowl alike, the animal specialist has had to figure out how to limit and control disease, and in some cases how to put out the fire of bird pandemics – for example, of Avian Influenza – or disease outbreaks among people.

In essence, erring on the side of erring, in the absence of clear evidence that overuse of antibiotics was dangerous, regulators and poultry-raisers did nothing until the 2000s to limit this input, if calling for some regulatory oversight. Of course, this is all too often the case with assessment of technology: we discover dangers late in our investigations, while in the meantime damage to public health has been done, and in the case of antibiotics the danger has become drug-resistant bacteria.

Foodborne diseases can be spread, of course, through forces – and fingers – outside of the factory-farm industry – for example, in restaurants or at home through unsanitary food preparation practices. There are simple practices to prevent these diseases – washing one's hands, knives and cutting boards carefully and regularly, refrigerating properly and using one's nose to avoid spoiled products. But perhaps here, too, we have become inured to the danger, egged on by the belief that food production is safe, that producers are vigilant, and that our chickens are happy, healthy and fine. We see vacuum-packed meat products in display cases, sitting row after row in bloodless boxes of muscle, and we think of gleaming, modern food factories that deliver fresh foods to the supermarket or butcher every day.

Government officials and food producers have also all too often pushed the onus for safety onto the consumer, informing her or him of this responsibility in a variety of brochures and public service announcements.[42] Admittedly, food handlers and consumers share some blame for poor practices. Public health announcements regularly proselytize about proper preparation and cooking, including how to avoid cross-contamination. Yet a recent study showed that even restaurant managers lack basic food-safety knowledge about chicken. "Forty percent of managers acknowledged that they never, rarely, or only sometimes designated certain cutting boards for raw meat (including chicken), [o]ne-third of managers said that they did not wash and rinse surfaces before sanitizing them," and over half never used thermometers to check final internal temperature of the meat. They didn't know the temperature to cook raw chicken so that it was safe to eat.[43]

In distinction from the EU where the practice is banned, in the US manufacturers of birds use a chlorine wash to kill microorganisms, rather than adopting other approaches to raising the broilers in the first place, that are less likely to be rife with bacteria problems – for example, more space for each fowl. Already in the 1950s, the industry had developed harsh chemical and biochemical means of dealing with the pressure of production, transport and sale of the industrial chicken that foreshadowed today's entirely physicochemical treatment of birds with baths, chlorine, antibiotics and the like. A veterinarian mentioned in 1957 new ways of prolonging storage life that included low holding temperatures, and chilling at just under freezing rather than 38 °F, with immersion in brine a possibility – or even in propylene glycol (anti-freeze!).[44] Antibiotics were also approved by the FDA in the 1950s for use in poultry chill water, with a maximum allowable tolerance of 7 ppm. Atkinson was certain they would substantially extend the storage life of poultry.[45]

Washing the chickens in a strong chlorine solution (20-50 parts per million of chlorine) is still considered in the US a "cost-effective method of killing any microorganisms on the surface of the bird, particularly bacteria," especially during slaughter and evisceration. But because of this practice, US chicken has been banned in the EU since 1997, not because the treatment itself is dangerous for humans, but because the EU Scientific Committee on Veterinary Measures worried that "heavily soiled birds may not be sufficiently disinfected, and that relying on chlorine washing could lead to poorer hygiene standards overall." Instead, EU officials focus on continually improving hygiene standards at all steps of processing, not on finding more chemical inputs – and more potential spots of hygiene failure. Since in the US there are no poultry welfare standards, the process is common. Indeed, according to the Humane Society of the United States, inhumane and unsanitary practices are rampant in the absence of animal welfare regulation.[46]

The dangers of *Salmonella*, *E. coli* and other dangerous bacteria are real, of course. *Salmonella* has long been a scourge of the poultry industry – for example, in the mid-1960s – with over 100 species of the organism, acute and chronic, that in chicks spread rapidly with high and sudden mortality. Survivors were often stunted and "unthrifty," with chronic diarrhea. Making matters worse, *Salmonella* lived quite a while in droppings even after fowl recovery, and spread to eggs. Even more serious, there had been a marked increase of the infections in humans – and no doubt even more unreported cases – that spread in food, in particular egg and poultry products. While mortality rates were low, the widespread incidence and misery of the patients in any event demanded vigorous and immediate control. One specialist called for compulsory testing of poultry feeds, prohibition of sale of any food with even suspected infection, and destruction of eggs and poultry as required.[47]

Antibiotics have been used excessively since the 1950s, not long after their introduction to the public to fight a series of different infections, so much a wonder drug that no one seemed to question the danger of overuse, and once that danger was clear, meat-makers still overused them. In the poultry industry, they have been overused because they cheaply accelerate growth. They were added to feed like toppings on a bird sundae. In the 1950s, about a quarter of the 2 million pounds of antibiotics produced in the US went to livestock feed, and by 1960 poultry ate 1.2 million pounds of wonderfeed per year. By the 1970s, many specialists were worried about the great danger of drug-resistant bacteria, although at the feed troughs of industrial scientists the argument centered not on stopping the practice but managing it by finding better drugs and smaller doses. In 1979, more than 40 percent of the antibacterials produced in the United States "were used as animal feed additives and for other nonhuman purposes. Nearly 100 percent of poultry, 90 percent of swine and veal calves, and 60 percent of cattle receive antibacterial feed supplementation. About 70 percent of U.S. beef by carcass weight comes from cattle that have received weight-promoting feed supplements." By the late 1990s, animals ate one half of the 50 million pounds of antibiotics produced.[48]

Eventually, in an atmosphere of growing understanding that overuse of the drugs would lead to drug-resistant bacteria and put humans at risk for "superbugs," regulations appeared. In 2006, the EU banned the practice of giving meat animals "subtherapeutic" doses of antibiotics in animal food and water. Denmark, the world's largest exporter of pork, further restricted use of antibiotics for growth promotion and routine prevention of diseases caused by overcrowded and unsanitary feedlot conditions. Danish researchers subsequently documented a decrease in antibiotic resistance in Danish livestock, yet no adverse effects on

Danish pork production levels.[49] Subtherapeutic antibiotics (STAs) can promote growth, particularly in poultry and hogs, by improving nutrient absorption and by depressing the growth of organisms within the animals that compete for nutrients, thereby increasing feed efficiency. But many drugs used to treat animals are the same as, or similar to, drugs used for human health care, and many specialists worry that the use of STAs in animals will again promote development of drug-resistant bacteria that could pass from animals to humans, and thus pose a danger to human health.

In the US, regulatory efforts often are slowed and weakened by the successful efforts of lobbyists who represent business interests. They convince regulators that new measures are too costly and unnecessary, and that self-regulation will work.[50] As a result, according to the US FDA, the use of antimicrobials continues to increase, although at a slower rate, with sales and distribution of all antimicrobials up 1 percent from 2014 through 2015, tying for the lowest annual increase since 2009, although the percentage of antimicrobials that are considered medically important in human medicine increased by 2 percent from 2014 through 2015.[51] Unfortunately, the FDA's requirement to report on this usage was introduced only in 2008.

In the US, precisely because of the power of members of Congress and senators from agricultural states to downplay safety concerns associated with STAs, and instead to favor the calls for help from agribusinesses to keep the current system intact, progress in restricting antibiotic use in livestock has been slow. The Preservation of Antibiotics for Medical Treatment Act (PAMTA), a bill that would phase out the use in animal feed of eight classes of medically important antibiotics, has repeatedly died in Congress.[52] The FDA issued guidance documents in 2016 for industry voluntarily no longer to sell their antibiotic products for "growth promotion," yet other identical antibiotic feed additives in the same dose range are still allowed. In 2014, the White House released a National Strategy to Combat Antibiotic-Resistant Bacteria, but the Strategy failed to set national targets to reduce antibiotic use in animal agriculture. The authors of one study urge public action, if not boycotts, to end these practices.[53] Indeed, this pressure has worked, with a number of leading fast-food chains having announced they will no longer sell food products from meat producers using STAs.

A number of studies point to the overuse of antimicrobials in the meat animal industry. For industry representatives, it would seem that these drugs are simply another good input to keep costs down by accelerating fattening and preventing bird loss. Yet, as Nataliya Roth et al. write, "The increase in antibiotic resistance is a global concern for human and animal health. Resistant microorganisms can spread between food-producing animals and humans." Further, producers and government officials have been inconsistent even in evaluating the problem. Roth et al. studied the type and amount of antibiotics used in poultry production, and the level of antibiotic resistance in *E. coli*. They examined national monitoring programs and research studies conducted in the US, China, Brazil and countries of the EU – Poland, the United Kingdom, Germany, France and Spain. They noted significant challenges in determining the risks because of "the absence of a harmonized approach in the monitoring of antibiotics per animal species and the evaluation of resistances using the same methodology. There is no public long-term quantitative data available targeting the amount of antibiotics used in poultry, with the exception of France." The data from France indicated that decreased use of tetracyclines led to a reduction in the detected resistance rates. The data also revealed that "the average resistance rates in *E. coli* to representatives of these antibiotic classes are higher than 40% in all countries."[54]

The World Health Organization has called antimicrobial resistance "an increasingly serious threat to global public health that requires action across all government sectors and society." If antibiotics administered to animals marginally improve growth rates and prevent infections, then there is growing evidence that antibiotic resistance in humans is promoted by the widespread use of nontherapeutic antibiotics in animals. Resistant bacteria are transmitted to humans through direct contact with animals, by exposure to animal manure, through consumption of undercooked meat, and through contact with uncooked meat or surfaces meat has touched. The costs are substantial and real, not only in terms of likely deaths from infections that cannot be treated. According to the Infectious Diseases Society of America, longer, more expensive hospital stays for treating antibiotic resistance cost the US health care sector an estimated $21 to $34 billion and 8 million additional hospital days annually.[55]

McKenna notes how the rapid transition to an antibiotic-based industry came about because it gave consumers cheap meat, while scientists and regulators, enchanted by modern technology, found little to object to at the early stages of what I call the chemicalization of birds that today consume four times the amount of antibiotics taken by humans. If the EU was quick to regulate and limit antibiotics, then the US considered controlling amounts of antibiotics only decades later. As a result, according to McKenna, there are hundreds of thousands of excess deaths that occur annually because of antibiotic-resistant bacteria, and the threat of much greater numbers – millions of people – exists by mid-century.[56] Ultimately, producers themselves began to scale back on antibiotic use, including Perdue Foods, which in 2007 stopped using growth-promoting antibiotics (although it announced this only in 2014), and had turned to vaccination and other methods to avoid illnesses, while a number of restaurant and food chains – Chik-Fil-A, McDonald's, Subway, Costco and Wal-Mart – joined the rejection of antibiotics. McKenna says that the decisions were practical more than anything else, because consumers were increasingly embracing antibiotic-free products, despite higher prices.[57]

The problem is global. Everywhere one hears a squawk, there is a diseased chicken who may be resistant to antibiotics. One study examined the incidence of *Campylobacter* in chickens in Malaysia, where an increasingly vertically integrated industry has taken hold and meat consumption is about 45 kg per capita. Researchers examined ten broiler farms and found *Campylobacter* in nine of them. Virtually all strains were resistant to such drugs as tetracycline, streptomycin, kanamycin and ampicillin.[58]

The problems seem insurmountable in industrial conditions. A 2015 review article documented the extent of *Campylobacter* in poultry (chickens, turkeys, ducks and geese) and wild birds, with prevalence rates, especially in slaughter-age conventional broiler flocks, as high as 100 percent on some farms worldwide. In spite of extensive research in the previous two decades, the authors noted that many gaps in knowledge remained and "effective intervention strategies" for the control of *Campylobacter* were still lacking.[59] As is the case with other bacteria, the use of antibiotics to prevent and control this bacterium led to antibiotic-resistant *Campylobacter* that can be transmitted to humans via contaminated meat. Two classes of antibiotics were a major concern for public health: in the mid-1990s, the FDA licensed two fluoroquinolones (sarafloxacin and enrofloxacin) for treatment of respiratory diseases in poultry, both of which now struggle with resistant bacteria.[60]

The chemical additives – antimicrobials – are of serious concern to public health given that they enter the human diet and also can be defecated into the land and waters. This has become a greater and greater problem in light of the tens of billions of fowl raised annually throughout the globe, most intended for delivery to distant markets where they may have a significant impact on public health. Distant trade and distant consumption have made it challenging to ensure food safety and public health, and to avoid avian disease epidemics.

Broiler Diseases: A New Specialization and Always New Diseases

Because it has been engineered to gain weight and move very little in a controlled environment, the broiler is at great risk for disease and for spreading bacteria. Therefore, the modern technobird must consume antibiotics and be vaccinated to limit the spread of disease between birds and between birds and humans. The spread of disease to humans may occur through food or directly from the animals to people. According to the EFSA, foodborne zoonotic (that is, between animals and people) diseases "are caused by consumption of food or water contaminated by such pathogenic microorganisms as bacteria, viruses and parasites. They enter the body through the gastrointestinal tract where the first symptoms often occur." The EFSA explanation continues, "The risks of contamination are present from farm to fork and require prevention and control throughout the food chain."[61] The severity of these diseases in humans varies from mild to life-threatening. When disease spreads rapidly among birds in factory-farm conditions, it may wipe out entire flocks or lead to the requirement that thousands of animals be killed ("culled"). This is a new and frightening industrial landscape that exists, once again, across the globe, and it includes dangerous outbreaks of Avian Influenza.

Scientists have been working to limit disease in poultry for decades, and yet avian medicine itself is a relatively new field. In the 1930s, few claimed specialization in the area, and few poultrymen turned to the local veterinarian for help in solving problems. But such institutions and agencies as the American Veterinary Medical Association, the US Livestock Sanitary Association, and the Poultry Science Association, and individuals associated with them, together with local organizations and government bureaus recognized the importance of study and pushed the development of the specialty. Even if diseases known to cause mortality in chickens had been studied before 1900, much of the knowledge of them and their treatment came from subsequent research.

Of course, specialists have long been aware of diseases that ravage chicken flocks. In the nineteenth century, Louis Pasteur, who modernized principles of vaccination, studied an epidemic that was destroying France's silkworms to arrive at the germ theory of diseases – that they were caused by bacteria. At Emperor Napoleon III's request, Pasteur further studied how to prevent wine spoilage. He found that by heating the wine to 50–60 °C (120–140 °F), he could kill any microbes that might cause contamination, a process now known universally as pasteurization. He developed vaccines for rabies and anthrax, and produced, after extensive research in the 1870s, a vaccine for chicken cholera, giving rise to the field of immunology. Pasteur served the chicken – as chickens served public health as the object of his experimental research to test vaccines. In Pasteur's mausoleum, you encounter these things: mosaics of blossoming plants and grapevines that allude to Pasteur's studies of fermentation; sheep grazing in a pasture; mulberry trees with silkworms and moths. And chickens. Pasteur had isolated *Pasteurella multocida* to grow it in a culture, but he and his assistant had allowed the sample to spoil and die. When he used this weakened sample purposely to infect chickens for further tests, he learned that he had not produced the disease in them, but immunity. Pasteur's chickens, unlike Pasteur himself, continue to hunt, peck and dance about on the walls of his crypt.

Avian Influenza was first reported in Northern Italy in 1878.[62] In 1880, Rivolta and Delprato showed it to be different from fowl cholera and called it *Typhus exudatious gallinarum*. In 1901, Centanni and Savonuzzi determined that fowl plague was caused by a virus. The highly contagious disease with high mortality continually caused massive outbreaks in poultry, including two outbreaks in the United States

(1924 and 1929). In 1955, scientists determined that the virus causing "Fowl Plague" was a type-A influenza virus.[63] (Types B and C affect only humans. One concern was whether type-A influenza is zoonotic.) In 1981, at the First International Symposium on Avian Influenza, the disease was named Highly Pathogenic Avian Influenza (HPAI).[64]

At the turn of the century, studies included *Investigations Concerning Infectious Diseases Among Poultry* (1895); a bulletin on fowl typhoid published in Rhode Island in 1902 by Cooper Curtice; a report on avian tuberculosis in Oregon by Pernot in 1900; and a variety of other studies.[65] Yet the field remained isolated and small. At the turn of the last century, of 48 Experiment Stations in the United States and Canada, only slightly more than half were engaged in the investigation of poultry problems, leaving 23 which were carrying on no poultry works.[66] According to specialists in the early years, investigations into disease remained rudimentary and preliminary, with "all lines of investigation in poultry diseases . . . profitable. Surveying the field of poultry diseases for the first time one is astounded that so little really final work has been accomplished." While scientists understood causality, the questions of origin, habitat and how to kill epizootic diseases "before they attack the flock," and with which weapons to do so, remained unclear.[67] One specialist posited that the reason for the lack of systematic study was a failure to appreciate the "importance and extent" of the industry, likely because of the fact that the individual units of production were "generally small and widely scattered."[68]

Beginning in the 1900s and 1910s, publications appeared that focused on this area of medical research, among them *Veterinary Bulletin*, the forerunner of *Veterinary Medicine*. In 1911, Raymond Pearl, with F. M. Surface and M. R. Curtis of the Maine Agricultural Experiment Station, published a 216-page Experiment Station bulletin on diseases of poultry, which seven years later was revised and published as a textbook. The specialization took off in the 1920s with dozens of veterinarian researchers entering the field of poultry diseases, with the rise of centralized diagnostic laboratories, in the US supported by agricultural experiment stations, veterinary schools, and state and/or county livestock control agencies. Three of the earliest of these were started from about 1917 to 1920 at Kansas State College; at University of California in the egg capital of Petaluma; and at the Western

Washington Experiment Station in Puyallup. By the 1950s, the poultry industry required "avian practitioners" to handle the large number of fowl – and to protect the investment of businesses. Special books and entire issues of journals now focused on the subject. According to a survey completed by the USDA in the early 1980s, by that time 93 state-supported poultry diagnostic laboratories had opened. Yet fewer than 20 percent of these facilities were sufficiently staffed and equipped for diagnosis of foreign and domestic diseases that required special technical equipment.[69]

The transformation of the bird into a packaged commodity came with a series of risks. The speed of growth, slaughter and evisceration, and the distance of transport, put significant health and safety pressures on the chicken industry. In spite of great progress in addressing public health concerns in the post-war years, one veterinarian worried precisely about transport and production revolutions as putting great stress on the chicken as a vector of disease. Some poultry continued to be delivered as so-called New York-dressed, ready for evisceration in the hotel or restaurant kitchen or in the housewife's home. These carcasses might sit for days or longer frozen or semi-frozen, at risk of contamination and spoilage, while those processed at high speeds and high volumes required better inspection, and the majority of plants and products had not been opened to such inspection. Buildings, equipment, operating procedures, waste disposal facilities and refrigeration practices were rapidly changing, and old theories had given way to modern practices (poultry carcasses *could be* and were safely suspended from a shackle or cut up on an impervious table or belt, in a manner that was faster and much more sanitary than the old "meatblock technique").[70]

In the late 1950s, most specialists had a rather positive view of US food safety – there had been remarkably low mortality and morbidity rates in the US recently, and less danger of contracting a foodborne disease than in almost any other part of the world. Yet they worried about recent disease outbreaks with thousands of cases of illness, and the thousands more of unreported cases. The problems were connected with salmonellosis, botulism, staphylococcal intoxication and toxic chemical agents, with botulism rare and usually coming from home canning.[71] The warning signs of disease pollution and the need for new inspection paradigms have been present for a long time, for over a century, but became a worldwide emergency with the rise of the supermarket, the factory farm, and the creation of distant markets even thousands of kilometers from the site of production and slaughter of broilers.

A specialist in the US Public Health Service (PHS), a doctor of veterinary medicine and a poultry specialist, James Lieberman, worried openly in 1954 about how to ensure wholesomeness of chicken products for the consumer, given the rapid expansion of poultry farming and the accompanying changes in processing, storing, packaging, transport and sales. Lieberman believed large-scale food processing required government-intervention. He stressed that inspection was not regulation for the sake of regulation. But given these changes in processing and transport, "let the purchaser beware" could no longer be applied. While poultry and poultry products were nutritious when wholesome, they also might carry diseases that were a public health concern, especially since fowl was a king of the animal "reservoir" of disease organisms.[72]

A large number of outbreaks of disease were due not only to foods like chicken meat, but to how the animals were handled. Yet, in 1954, US federal inspections involved a two-phase-only voluntary program that covered, at most, 20 percent of poultry, and then only poultry that was shipped in interstate commerce. States and locales were left to pick up the slack, and thus, Lieberman argued, national standards to control the hazards, inspect for wholesomeness and sanitation, and enforce proper health standards were required. In 1952, the PHS proposed a poultry hygiene standard that brought industry on board, the latter wanting uniform practices – and also likely hoping to ensure that, when finally adopted, the government standards were not too onerous. The standards were obvious: no rodents or insects, proper disposal of refuse, employees who were properly dressed and also clean, and ante- and post-mortem inspection of poultry.[73]

In 1985, as the US industry expanded ever more rapidly, as the number of birds being raised, pushed into industrial settings, slaughtered by conveyor-belt method, dipped into various bacteria killers, chilled, chopped into pieces and wrapped, and transported to markets even thousands of kilometers away continued to grow, a committee of the National Research Council's Food and Nutrition Board (FNB) completed a report on the scientific basis of the USDA meat and poultry inspection programs. The task was to carry out a risk–benefit analysis on whether bird-by-bird inspection could be abandoned for statistical, "less-than-continuous" procedures, if based on the methods, approaches and concerns of public health, food microbiology, toxicology, risk assessment, risk management, veterinary pathology, poultry inspection technology, biostatistics and epidemiology. Based on a thorough review of literature, inspections of two poultry production plants, and interviews with federal veterinarians, poultry producers and consumer representatives, the committee determined that "in general it found that it is not possible to determine from existing data whether current inspection programs actually fulfill their goal of protecting the public health." The committee therefore considered whether to recommend a move to a newly proposed, less-than-continuous post-mortem inspection system. It concluded that no such changes could be proposed until justified by a detailed risk analysis of the public health risks involved. It further recommended that the department's Food Safety Inspection Service (FSIS) "establish a risk-assessment program that applied formal risk-assessment procedures to assist in planning and evaluating all phases of poultry production in which hazards to public health might occur."[74]

A revised system of inspection clearly was needed, since each year some 1 in 50 Americans suffered from gastroenteric disease due to *Salmonella* and *Campylobacter j.* The diseases are particularly dangerous for the very young, the very old and people with immunosuppressive disease. But already in 1985, some 4 billion broilers passed through the system annually. Largely, the system ensured that these "vehicles of bacterial and chemical contaminants" were for the most part "wholesome."[75] The committee members determined that the system of continuous inspection offered little opportunity to detect or control the most significant health risks associated with broilers, especially since it could not detect pathogens at each stage of production and slaughter. Indeed, microorganisms pathogenic to humans were present in broilers at time of slaughter and retail sale. The determination was to expand the traditional focus from slaughter to include such other potential sources of poultry-related hazards as production, preparation and handling, and cooking.[76]

Was there success? The FSIS celebrates its inspection system as being "science-based," yet a major reform happened only eight years later in response to an unprecedented outbreak of *E. coli* O157:H7 in the Pacific Northwest in 1993, the latest of several, that caused 400 illnesses and 4 deaths – and public outrage demanding safer food products. Until that time, FSIS inspection was "largely organoleptic (relying on sight, touch, and smell)." In 1996, FSIS issued a landmark change for Pathogen Reduction / HACCP Systems that focused on the prevention and reduction of microbial pathogens on raw products that can cause illness, clarified the roles of government and industry, with the latter "accountable for producing safe food," and the former "for setting appropriate food safety standards, maintaining vigorous inspection oversight to ensure those standards are met, and maintaining a strong regulatory enforcement program to deal with noncompliance." It seems to have been effective, according to the CDC, in promoting an "overall decline in bacterial foodborne illnesses since 1999."[77]

Yet outbreaks of foodborne illness continue year by year, and recalls of tainted food products have become almost an accepted part of daily life. Finally, in response, in 2011 President Barack Obama issued Executive Order 13563 that required the USDA to update its inspection regulations. The USDA issued in early 2012 a 379-page proposed New Poultry Inspection System (NPIS) for chicken and turkey slaughterhouses. In one concession to consumers, it stipulated that production lines should run no faster than 140 birds per minute, although industry pushed for higher speeds. To great fanfare, in January 2014, after two years of comment and discussion, the FSIS announced a critical new program to make mass-produced chicken and turkey products safer, in particular by giving poultry companies an NPIS to control *Salmonella* and *Campylobacter*, and to prevent up to 5,000 foodborne illnesses each year, FSIS claimed, through a "science-based inspection system that positions food safety inspectors throughout poultry facilities in a smarter way." It replaced a system that dated to 1957, and imposed stricter requirements on the poultry industry to prevent contamination in the first place, rather than addressing contamination after it occurs, through the industry's own microbiological testing at two points in the production process. FSIS also introduced the "optional NPIS" through which poultry companies first sorted poultry for "quality defects" before presenting it to FSIS inspectors. FSIS claimed this would allow a focus less on routine quality-assurance tasks and more on strategies that are proven to strengthen food safety – for example, being able to more frequently remove birds from the evisceration line for close food safety examinations, take samples for testing, check plant sanitation, verify compliance with food safety plans, observe live birds for signs of disease or mistreatment, and ensuring plants are meeting all applicable regulations.[78]

First results showed, in fact, a worsening of conditions, and demonstrated that allowing the regulated to regulate themselves will only result in more danger to the public. As noted by Food & Water Watch, when, after more than a one-year delay, on January 23, 2018, the FSIS finally posted the results of the agency's regulatory sampling for *Salmonella* in the nation's poultry slaughter plants, the results revealed that the plants that converted to the NPIS failed the agency's *Salmonella* performance standard. And how could it be otherwise, since the NPIS removed most of the USDA inspectors from the slaughter line and turned inspection responsibilities over to company employees to perform, with perhaps two to three times higher contamination rates in many of the slaughter plants than before? Wenonah Hauter, Executive Director of Food & Water Watch, said: "It's clear that this privatized inspection system that was hyped as an improvement to food safety certainly isn't. . . . In fact, it has higher rates of contamination than slaughterhouses with more government oversight."[79]

Indeed, the factory-food industry worldwide makes a fail-safe system impossible to achieve. Nearly every day, in the US alone, the FSIS website lists one or two product recalls of 50,000, 70,000 or 200,000 pounds of meat, for contamination by bacteria or foreign matter – for example 4.5 million pounds of Tyson chicken nuggets in 2019 with rubber pieces in them, or 2.5 million pounds of breaded chicken in 2017 for mislabeling. And so on. Indeed, it is likely impossible for the FSIS and its 9,000 employees to inspect the food industry with its 6,200 facilities nationwide in order to protect public health. In Arkansas alone, with its 1 billion chickens, there are 459 employees, not all of whom work in the field, responsible for 107 facilities in 53 of the state's 75 counties. There are 2,400 broiler farms. Altogether, Arkansas poultry employs directly roughly 40,000 people.[80]

Food Safety and Machined Chickens in an Era of Avian Influenza

The challenges of understanding diseases increased with the rise of factory farms, the international and transboundary trade relationships that opened new vectors for the spread of disease, and the conflict between traditional methods of raising birds and industrial ones, with traditional farm owners often unwilling or unable to adopt biosecurity measures, let alone to slaughter entire flocks to prevent further spread of disease. The outbreaks are sudden and the impacts great. The risks include, once again, animal-to-human infection. The disruptions to life at home and trade abroad have been significant, especially with the rise of HPAI.

HPAI has occurred in most parts of the world. From 1959 to 1995, the emergence of HPAI viruses was recorded on 15 occasions, but losses were minimal. In contrast, between 1996 and 2008, HPAI viruses emerged at least 11 times, and 4 of these outbreaks involved many millions of birds – and thousands of humans, hundreds of whom died. The first cases of human infection with Avian Influenza A (H5N1) were identified in 1997 in Hong Kong. The virus infected 18 persons and caused 6 deaths. Specialists believe that the immediate culling of around 1.5 million poultry in Hong Kong averted a larger outbreak in humans. In 2003, an outbreak of H7N7 Avian Influenza in the Netherlands caused the death of 1 veterinarian, and sickness in 83 other people. In December 2003, a major epidemic of HPAI hit chicken farms near Seoul, Korea. Eventually the authorities killed hundreds of thousands of chickens and ducks. By early 2004, outbreaks of the same virus hit Cambodia, China, Hong Kong, Japan and Thailand. The 2003–4 outbreaks in Vietnam led to human fatalities. Beginning in October, hospitals in Hanoi and surrounding provinces admitted 14 persons with severe respiratory illness, mostly children, and 11 of the children and the 1 adult died. The virus spread rapidly and had a mortality rate in chickens approaching 100 percent, leading to the deaths of 40,000 chickens and the culling of 30,000 more. Another outbreak in South Korea in late 2006 and early 2007 resulted in the deaths and culling of thousands more birds.[81] Next an epizootic of HPAI due to an H5N1 virus spread throughout Asia and into Europe and Africa, affecting over 60 countries and causing the loss of hundreds of millions of birds.[82] All sectors of the poultry population were affected. Free-range commercial ducks, village poultry, live-bird markets and fighting cocks were significant in the spread of the virus. According to Alexander and Brown, the role of wild birds has been extensively debated but it is likely that both wild birds and domestic poultry are responsible for its spread.[83]

These outbreaks are frightening beyond the risk to public health. They damage consumer confidence, contribute to loss of markets and shortages, disrupt trade flows, and in fact contribute to restructuring trade flows: the outbreaks contributed to Brazil's emergence as the world's largest supplier of frozen raw chicken products, and they led poultry industries in Southeast Asia to convert production from unprepared to prepared poultry meat for export.[84] A world crisis involving the broiler has been the result.

Even with "biosecurity" measures in place (segregation, cleaning, disinfection, slaughter, disposal and isolation), the environmental circumstances of HPAI make it challenging to contain. Making control more difficult is the fact that:

> live infected domestic poultry can produce virus for several days or weeks without clear clinical signs. Infected domestic birds are the most dangerous source of virus and inanimate objects (fomites) contaminated with secretions (in particular feces) from infected birds are the next most dangerous source of virus and air-borne spread is not significant. The disease is mostly spread by the actions of man, moving either infected birds or contaminated materials.[85]

Even with Avian disease veterinarians present and international controls and quarantines in place, from 2003 to 2007, as we have seen, hundreds of millions of birds, wild and domestic, died from the Avian flu or were culled. Across the globe, governments tried to control the spread of Avian Influenza by imposing import bans on poultry products originating from HPAI-affected areas. As biosecurity is practiced mainly on large-scale farms, small flocks and live-bird markets that are common in Southeast Asia often lack the controls and practices necessary to prevent infections, and thus may be "the main reservoir of the virus and live markets are thought to be the principal source of contagions and outbreaks."[86]

The impact of HPAI may hit as hard in the US as it has in Europe and China, even with biosecurity measures, as can – and often do – *E. coli* and *Salmonella*. Prevention and treatment are problems since "current animal identification and meat product labeling practices make it difficult or impossible to trace infections to the source." On top of this, factory farms are often exempt from participating in "public health monitoring, disease reporting, and surveillance programs."[87] And let us not forget the risk to migrant and visiting workers, many of whom are undocumented. Obviously, air emissions are a crucial concern and "children, the elderly, individuals with chronic or acute pulmonary or heart disorders – are at particular risk." Respiratory and neurobehavioral diseases, symptoms and impaired functions have been documented.[88]

The experience of Chinese chickens indicates the rapidly changing nature of raising broilers in factory farms. The Chinese poultry industry is struggling with falling consumer confidence and a decline in consumption related to Avian Influenza and other scandals. A turn to processed animals – parts, nuggets and so on – has not generated returns because Chinese consumers have traditionally preferred to butcher the entire bird at home. Export markets also are contracting because of safety and flu problems. The international plan to stop the virus's spread and lower the risk of mutations was to track and eradicate outbreaks at the first moment, with local officials culling an afflicted flock. But Chinese chicken farmers instead tried to prevent H5N1 by treating their birds with Tamiflu, an antiviral drug that has been effective against H5N1. According to one source, "As a result, it became that much more difficult for health officials to track H5N1 outbreaks because Tamiflu-dosed chickens could still get infected and spread the virus but without showing the symptoms that would set off medical alarm bells."[89]

The global broiler industry, with its controlled environments and plastic surgeries, with its antibiotic treatments and cellophane packing, hides from public view a series of paradoxes, like the broilers themselves. They are animals, yet bloodless machines; sentient, yet motionless; cacophonous, yet silent; chickens, yet processed industrial objects. The industry is highly centralized and dominated by a few producers, yet claims itself to be an amalgam of small farmers. It contends that it operates safely and produces meat cheaply, yet it faces constant public health and animal welfare problems whose costs are passed along to the consumer. Today's consumers get chicken in clear-wrapped dishes, *sans* feathers and seemingly even follicles, and certainly without feet or head. Cut into manageable cooking pieces, or sold in assortments of legs, wings and breast, boneless or with bones, skinless or with skin, chicken hardly resembles a once living thing. In the absence of well-funded and fail-safe state and international food inspection programs and interventions, there are also growing dangers of foodborne bacteria. Labels at the checkout counter never indicate these flesh-and-blood realities. La Mettrie's vision has been realized in dystopian form.

4

Shit and Feathers

> You can throw shit, sling shit, catch shit, or duck when the shit hits the fan. You can take a shit, give a shit, keep shit or serve shit on a shingle. You can find yourself in deep shit, or be happier than a pig in shit.
>
> – George Carlin, stand-up comedian

How much manure does it "cost" to produce 2 kilograms of chicken flesh? The answer is 1 kilogram. And we shall have it. Billions of tons. Deposited on the earth. Dried. Wet. Stored. Stored poorly. Mostly stored poorly – if at all. From Brazil to China to Europe.

Just over 20 years ago, in 1999, the US General Accounting Office (GAO) sounded the alarm about the vast quantities of agricultural animal waste our farms produce: in that year, 130 times more than the human waste produced – at roughly 5 tons per person annually – and often from CAFO operations with hundreds of thousands of animals. Most CAFOs produced more waste than the entire nearby town or city. All of this shit was the result not only of the expansion, but of the concentration in one place, of animal production for livestock and poultry – and without any spot to put it, or any thought what to do with it. In the broiler sector, operations with 100,000 or more birds increased from about 70 percent of national sales in 1974 to about 97 percent in

9 *Avian Influenza threatens the international bird trade and requires heightened efforts at biosecurity. Here, a man in hazmat dress plucks feathers from bird corpses for eventual testing for influenza.*

1992. The waste threatened "surface water and groundwater quality in the event of waste spills, leakage from waste storage facilities, and runoff from fields on which an excessive amount of waste has been applied as fertilizer." New waste management practices were needed, including alternative uses for waste, new means of treating waste, and improved methods of moving waste to cropland where it might be used as fertilizer. The GAO suggested limiting runoff by cementing and curbing animal confinement areas, or planting grassed buffers around these areas, or using storage tanks or retention ponds, although these storage ponds and lagoons have been shown to be only stopgap measures, and have been overwhelmed in storms by heavy rain. Other possibilities included using manure in feed mix, composting it, or as biomass for on-farm energy generation.[1] Although the animal-meat businesses rake in profits, the government offered them millions of dollars more to find a solution to the shit problem, encouraging them to seek grants to study various options. Meanwhile the Agricultural Research Service was spending $10-15 million annually on research and development (R and D) on waste runoff. But little has been accomplished, and the situation 20 years later is much worse.

Europe has made a bit more progress in handling this waste, but there, too, the shit is flowing deeper and faster. If, in the past, such residues as solid and liquid manure from cattle, pigs and chickens was used for fertilizer, then, with the onset of the industrialization of agriculture and massive centralized animal production, there is too little land for the application of waste within short distances of the sources. CAFOs and government organizations are therefore pursuing separation, drying, composting, anaerobic digestion and methanization.[2] Chickens and other industrial meat animals produce so much shit that data are incomplete about precisely how much manure there is. A survey of roughly 10-12 percent of the cattle, pig and poultry farms in France in 2008 indicated that most manure was applied to land, although anaerobic treatment and digestion were on the increase; this suggests that treatment was quite underused as an option, and the technology had not reached sufficient maturity.[3] The French, by the way, love their shit. *Merde!* French farmers and other protesters have made it a recent practice to dump it in front of government buildings and use spreaders to fling it against buildings.

The Chinese, who are pursuing biogas, have a real problem. Since the opening of the private agricultural sector in 1979, meat farming has grown rapidly, and by 2010, in the results of the first national pollution census, the government revealed that livestock production, which previously had largely been ignored in environmental management, was found to be responsible for 38 and 56 percent of total agricultural nitrogen and phosphorus non-point-source pollution, respectively.[4] To date, there has been little success in putting a dent in the massive, growing piles. But China may have found a solution: do a search for "chicken manure" on the website of the world's largest retailer and e-commerce company, Alibaba, and there are hundreds of machines – dehydraters, screw presses, cleaners, turners and other devices – available for quick purchase to dry and handle your waste, and also packages of chicken shit available for purchase.

Now, mostly, factory farms attempt to dry droppings for a rainy day, or spread it on cropland. Of course, dumping on fields has significant dangers, too. Animal waste runoff damages surface water and groundwater by introducing such pollutants as nutrients (including nitrogen and phosphorus), organic matter, sediments, pathogens (including bacteria and viruses), heavy metals, hormones, antibiotics and ammonia. The pollutants are transported by rainwater, snowmelt or irrigation water through or over land surfaces and are eventually deposited in rivers, lakes and coastal waters, or introduced into groundwater. They affect water quality and public health in several ways, such as contaminating drinking water supplies and killing fish.[5]

Manure is a problem in smaller operations as well. Investigation of waste management practices in Gaborone, Botswana, indicated that roughly 80 percent of chicken farmers disposed of manure and litter by giving it away to other farmers, 16 percent used it as a fertilizer on their own fields, and 4 percent of the respondents disposed of manure/litter in the landfills and other dumping sites for incineration or burning. Even with the small quantity of manure produced in 2012 in Gaborone – 20,441 tons – the farmers faced significant challenges in disposing of it, including poor transport, lack of farmers' knowledge of poultry waste management, and insufficient storage space. On top of shit, the farms had litter, feed, feathers, hatchery waste (empty shells, infertile eggs, dead embryos and late hatchlings), sludge and abattoir waste (offal, blood, feathers and condemned carcasses) to handle. Some farmers scraped the litter and manure off the floor using spades, shoveled it into empty 50-kg bags, and awaited collection, or once a year used a skid loader to move it on trucks or their own vehicles. The dumps and landfills were located far from poultry farmers. The result was compromised biosecurity – and massive outbreaks of houseflies and unpleasant odors that offended neighbors,[6] and no doubt a few farmers who could not be troubled to cart it away.

Several major studies have documented the extent and growing social, economic, and environmental impacts of the transition to factory farms for production of meat animals, concerning manure and other wastes. One Pew Memorial Trust study noted that, from 1970 to 2008, chicken consumption in the US alone grew from 40 lb per capita to 80 lb, surpassing pork in 1985 and beef in 1992. Of the 1.6 million farms growing chickens in 1950, only 2 percent remained, while sales jumped 1,400 percent to 8 billion birds by 2007. The industrial enterprises had no safe way to deal with all of the waste generated by these birds, and generally dumped it on fields, or carted it off to be dumped on someone else's fields. But, whether here or there, it washed off the land into local streams and rivers, and thence into bays and oceans, and in relation to chickens, especially in the Chesapeake Bay Region, around the Delmarva Peninsula. The nitrogen and phosphorus that fill the water lead to eutrophication: algae growth, loss of aquatic life, beach closures, shellfish contamination, lower oxygen levels. Sixty percent of the nitrogen (N) and 61 percent of the phosphorus (P) that enter the environment from agriculture comes from poultry operations.[7] The focus in this chapter is largely on the situation in the United States. But the scale of the problem in the US indicates, and the tardy and largely inadequate efforts to deal with it are instructive of the challenges faced by the growing number of broiler operations in other countries: volume of waste, runoff, absence of storage facilities, poorly developed alternatives to using manure such as biogas, and eutrophication of waterways from unmanageable and unmanaged quantities of N and P.

Factory Farms as Nature's Outhouse

At one time, specialists assumed the manure waste problems could be managed – through some as yet undiscovered technological sleight of hand. Or perhaps waste could be hidden in rural areas, or would just go away. Or perhaps people did not recognize any problem early on. US Department of Agriculture analysis in the 1970s reveals a sanguine attitude about the waste from these billions of birds, as compilers contended that "nearly all the offal and feathers and most of the blood from broiler processing are salvaged and rendered to be used as protein supplements in animal feeds," although they acknowledged that the nearly 3.5 billion pounds of byproducts – and water used to wash it away – created significant pollution problems, even if industry handling of byproducts was better than 15 years earlier. Unfortunately, a large number of the operations were dependent on municipal facilities for waste treatment – over 60 percent of them, and 7 percent had no waste treatment whatsoever. And, of the private facilities, 63 percent used lagoons of some sort to hold biological material, and about half of them combined aerobic–anaerobic systems. To meet appropriate environmental standards, the authors of one study concluded that the cost of using "the best available technology" would increase current levels of $10 to as much as $34 per 1,000 birds processed annually.[8]

Even before the rise of CAFOs, poultry wastes were a significant problem and taxed sewage systems and water treatment facilities everywhere that there were large farms and processing facilities. The history of industrial pollution from the nineteenth century reveals that producers rarely anticipated – and rarely could anticipate – the significant environmental and public health costs of production byproducts. And, even if they knew of dangers, they dumped the byproducts willy-nilly on-site or into streams or ponds to dilute noxious stuff or carry it away, and the significant dangers to the public were buried or ignored, like the wastes themselves, for decades to come. Certainly, public health officials knew of connections between polluted water sources and diseases like typhus and cholera, and they suspected that heavy metals had neurological effects. In the nineteenth century, municipalities struggled to secure clean water and to move waste from the cities in sewers. They began to regulate safety in mills, mines and meat-packing plants. In the early twentieth century, they understood that lead was an aggressive affront to human health, and banned it as an additive in gasoline, for example, already in the 1920s (although in the US only in the 1970s). They realized that tanning and other factories and workshops were centers of suffering and pollution. Yet with factory farms, even with the obvious knowledge that vast quantities of shit require special handling, the facility owners have done little, other than to spread the risk on nearby fields. For their parts, regulators did next to nothing, offering factory farms special dispensation for their sins and treating agricultural runoff as just a fact of business.

To put it another way, it is almost a rule that industrial growth – and recognition of accompanying pollution, safety and other "exogenous" costs – will precede the knowledge and ability to develop public health measures required in response. And it is a second rule that industrialists will pass on those costs downstream wherever possible. Take the case of the Georgia poultry industry, one of the birthplaces of integrated fowl production, located mostly in the northeast corner of the state. The industry took off in the mid-1930s and, by 1958 in Gainesville, Georgia, consisted of eight major plants processing an average of 270,000 birds daily and a maximum of 330,000 birds. Think of the eggshells from the chicks, the shit and urine from the birds, and the offal, feathers and blood those birds produced. The waste thus consists of shell sand (finely ground shells that are a source of limestone) and liquids (blood) that flowed into troughs to drums for storage. During processing, more waste is produced as the birds are mechanically picked, washed, cleaned, washed again, and finally removed from the chain. At strategic locations along the chain, workers dump feathers, offal, heads, feet and other scrap and waste products. Grease, solids and smells overwhelm the worker and the sewer system. Rendering plants nearby take the offal, feet, heads and scraps to be used as feed additives, pet food and fertilizers. While these waste products make money, there is little incentive to gather and control any more of it, and much of the waste – often carried by washing or waste water – flows into the sewer system. Meat animals require a lot of water to process, and each meat bird in Gainesville used over 3 gallons of water on average during processing. One early study concluded that about 50 lb (roughly 23 kg) of waste was produced per 1,000 birds,[9] but this feels like a significant undercount.

Prior to 1949, the Gainesville sewerage system consisted solely of a series of outfalls and laterals that discharged into the natural drainage area – into wetlands and waterways nearby. Of course, streams rapidly deteriorated as they filled with muck, guts, blood, feathers and wastewater. A new outfall was built that pumped waste into the Chattahoochee River drainage area. In other words, engineers worked by the mantra "the solution to pollution is dilution" and sent it downstream for further extensive impact on local bodies of water and land, on fisheries, on drinking water, on plant life. Engineers eventually realized the need for a large dam and reservoir to collect and treat all sewage and waste in a common plant; according to the plan, unfortunately, the reservoir was intended to handle only 60–85 percent of all wastes.[10]

Within a year of its completion, chickens (and their industry) had overwhelmed this treatment plant with grit and feathers, and with suspended and precipitating solids. Only in 1957, long after pollution problems had dictated the construction of the treatment plant, did the city pass an ordinance requiring all processors to install, operate and maintain satisfactory screening installations at their effluent points. But, at the end of that year, only 50 to 60 percent of the waste was being screened. Feathers were a particular problem, even though they were removed along the way by mechanical rake; grinders were too easily overloaded and stalled with matted, wet and wedged fathers. Pumps had problems with the sludge. Poultry waste did not follow expectations about when it would float or sink.[11] And today there is more of it, much more of it – billions of kilograms around the world.

The increasing scale and geographic concentration of the poultry industry has created a host of new environmental problems in connection with this waste, on a scale and extent several orders of magnitude larger than that faced by Gainesville municipal waterworks. In a number of high-production areas, the volume of poultry waste now exceeds the absorptive capacity of local and regional ecosystems, impairing the quality of surrounding waterways. On the Delmarva Peninsula, which produces more than 600 million chickens annually, the regional environment must contend with some 1.5 billion pounds (680,000 metric tons) of manure every year – more chicken shit than the waste load from a city of 4 million people. The growing question, as the mounds of manure grow in heaps and bounds, is how to deal with it and where to get rid of it. The shit, with its rich nitrogen (N) and phosphorus (P) content, has overwhelmed ecosystems. It would appear that the only remaining alternatives are to incinerate it, use it as biomass, or require other treatments. At one time, nuclear wizards suggested using peaceful nuclear explosions to create underground waste storage caverns. The day may come soon when such a solution for chicken shit storage may be advanced. For the time being, CAFOs might contract to use the Yucca Mountain nuclear waste facility in Nevada, destined for spent nuclear fuel, since that facility has not opened – and likely will not open – for nuclear waste.

One challenge to the shit disposal problem is vertical integration, with the producers passing off their responsibility onto contractors for waste disposal – and the contractors are clearly overwhelmed by the quantities of manure, their lack of comprehension about the laws and regulations concerning shit, and their financial inability to deal with it. If 95 percent of broiler producers in 1950 were independent, then only 5 years later, the number of independent growers had plummeted, accounting for only 10 percent of broiler production. There has been "a frenzy of consolidation" since the 1950s and 1960s – with larger entities coordinating production at each stage to avoid either overproduction or shortages, and buying feed, medicine, equipment and other supplies at bulk discounts, passing on those costs to the contractors who buy from them,[12] and they pretend that pollution is not their concern.

Another challenge to shit disposal in the US is regional concentration of the industry. A precipitous decline from 1.6 million to 27,000 farms has involved their concentration in states more friendly to still-unsolved and growing pollution problems, with relatively weaker law enforcement, and also weak in terms of unions. They include states across the US south: Texas, Arkansas, Alabama, Mississippi, Georgia and North Carolina. In the "broiler belt" from Delmarva to Eastern Texas, birds outnumber people as much as 400 to 1 in some regions. The accumulated waste has resulted in environmental and community challenges that are, in fact, a real catastrophe.[13] Roughly a third of US broiler farms have no cropland, few have sufficient cropland available nearby, and 60 percent of broiler litter is transported away but with no final resting place. Most farms that apply manure apply it to all of their land, including pasture. But farm size is the key here: larger broiler operations are more likely to remove all of their manure, and larger operations have higher land application rates.[14]

Because of their need for water, farms are often close to wetlands and rivers, and many factory farms are sited in or near fragile ecosystems, "such as on flood plains in North Carolina and over shallow drinking water aquifers in the Delmarva Peninsula and northeastern Arkansas." The facilities generally can have "substantial adverse affects on air, water, and soils." They use specially formulated feeds with "antibiotics, other antimicrobials, and hormones to prevent disease and induce rapid growth," but at the risk of proliferating antibiotic-resistant strains of bacteria.[15] There are extensive health risks also from exposure to various wastes, aerosols and so on – including in the manure, the air, the water – that can migrate a long way in the aquifer and affect drinking supplies of local communities, workers in processing, farmers, farm workers, and their families who "typically have more frequent and more concentrated exposures to chemical or infectious agents."[16] To be precise in such a messy subject, "Poultry is responsible for 44% of the total feces production [of farm animals] in the U.S., followed by cattle and swine."[17]

Crossing a Broiler's Ns and Ps

One of the most significant impacts on the environment from chicken shit is excessive quantities of N and P. Nitrogen and phosphorus wash into waterways, feeding algal blooms that deprive them and the species in them of oxygen. They lead to explosive growth of toxic algae and microorganisms that result in fish kills. Vast quantities of feces from different sources – cattle, swine, sheep, horses and poultry – have had an impact on every state, so much so that it can be measured in thousands of kilos of N and P by state and by square kilometer. Alabama, fiftieth in education and forty-fifth in health care in the US – according to funding, test scores and other criteria – seems to care more about chicken feces, because it allows its CAFOs to produce 114 million kg of N annually, or nearly 24 kg per person, while Iowa won the phosphorus battle with 154 million kg of P – nearly 50 kg per person annually.[18]

N and P overwhelm the capacity of the land to absorb it, carried in animal feces at concentrations 75 times "more concentrated than raw human sewage and more than 500 times more concentrated than the treated effluent from the average municipal wastewater treatment facility." Chemical pesticides and herbicides have entered the mix. Algal blooms have become widespread, creating dead zones "affecting an estimated 173,000 miles of US waterways." Animal facilities also contribute significantly to soil and sediment erosion. And when extreme weather events occur – and they are occurring with greater frequency – then "the risk, and quantity, of runoff" do also. These events include hurricanes and floods which overwhelm waste storage facilities whose runoffs carry "pesticides, heavy metals, and antibiotics and hormones" – from zinc and copper to antibiotics, from pesticides to nitrate – far and wide, with significant impact on public health, in concentrations far above the amounts allowed. Finally, ammonia emissions are another significant problem.[19]

Things are just as bad in Delmarva, with resulting devastation of Chesapeake Bay fish, crab and other ecologies. Studies showed no reduction in waste in the Delmarva region over the first decade of the twenty-first century, and growing spread of fish-killing dead zones in the Chesapeake Bay. Some regulations to limit sharply poultry litter spreading in Maryland have been introduced.[20] In the meantime, Tyson spokespeople stand *waste* deep in shit, claiming that manure and litter are valuable commodities, and that they have explained to their contractors how to manage them.[21] In 2009, Maryland and Delaware alone produced some 523 million chickens – 6 percent of the entire nation's production on less than 0.5 percent of its landmass. Assuming that 1,000 broilers produce roughly 81 cubic feet of litter, the chickens in the two states generated over 42 million cubic feet of litter. And the runoff of N and P into the Chesapeake continued unabated. The crab industry in particular has collapsed, with climate change and pollution leaving it without claws to stand on.

To assist in the growing problem of dealing with lots of shit, the EPA has published manuals on how to deal with it, after first defining standing rules and procedures to be met, necessary for factory farms to comply with the Clean Water Act. The EPA has determined that CAFOs must apply for a waste disposal permit unless they have received a "no potential to discharge" determination from the permitting authority, "where there is no potential for any CAFO manure, litter, or process wastewater to be added to waters of the United States under any circumstances or climatic condition." It identifies the requirements for handling of the manure and preparation of land prior to land application of manure, litter and process wastewater.[22]

From an environmental standpoint, the worst problem is a lagging regulatory response: the modern-day Clean Water Act that helped achieve "dramatic reductions in water pollution across the country" has not been vigorously applied to CAFOs, which are often not treated as "point sources" of pollution.[23] Although the law designates CAFOs as point sources, it also specifically exempts "agricultural stormwater" from the permit requirements. CAFOs, according to Pew Memorial Trust analysis, thus have "been given a unique but entirely ineffective dual regulatory status: regulated as point sources for waste in broiler houses and storage areas, but treated differently when that same waste runs off the cropland where it is applied."[24] We see this all too vividly when hurricanes dump vast quantities of rain on offal lagoons and shit-saturated fields, and the immediate neighborhoods and downstream waters and fields become filled with shit and offal – and dead fish. And floating carcasses of animals. Hurricane Florence in September 2018 killed 3.4 million broilers and thousands of pigs.[25]

Finally, because of expanding state and federal regulations, certified nutrient management plans (CNMPs) are coming into widespread use. For 2004–6, 62 percent of US hogs, 60 percent of broilers, and 49 percent of dairy cows were produced in operations that have CNMPs. But the amount of "nutrient applications" to fields (shit and other residues) will need to be reduced to comply with many of the plans. More cropland acres for manure application may be found by applying manure to more of a farm's cropland, by acquiring more cropland, removing manure from the farm and applying it to crops on other farms, or perhaps by incineration. In other words, the amount of manure continues to overwhelm the land, and broiler farms remain unprepared to handle their waste. The amount of land needed grows with each billion birds, and they do not allocate sufficient funds to acquire this land or remove the manure.[26] Broiler manure generally contains two to four times more nutrients than are contained in manure of other livestock, and 40 percent of US broiler production occurs on farms without any crop acreage for disposal by spreading.

One example will indicate the extent of the problem. In 2010, in the Lake Eucha and Spavinaw Creek watersheds in northeast Oklahoma, growers produced 44 million chickens in approximately 2,450 chicken houses. Up to this time, the growers spread the manure on cropland, but this has created algal blooms and degraded drinking water supplies. Voluntary methods, unsurprisingly, failed to help. Lake Eucha was created in 1952 by the Eucha Dam on the Spavinaw Creek and holds 100 million m^3 of storage water for the City of Tulsa, Oklahoma; Lake Spavinaw is the principal municipal water source for Tulsa. Tulsa and its Metropolitan Utility Authority sued to gain clean-up of the watershed and force the growers to adopt a workable plan to end the disgusting practice of allowing shit runoff into the water. The situation was so bad that water quality failed to improve as much as hoped, and legal wrangling continues against poultry companies.[27]

Similar legal actions are occurring across the nation. Iowa's largest municipal water utility provider sued a number of upstream drainage districts for excessive drinking-water nitrate levels caused by farmland runoff. The lawsuit was dismissed, with the judge ruling it a problem for the state legislatures to tackle. This was an unfortunate decision and a cop-out, because CAFO pollution goes beyond state borders and because citizens are drinking unsafe water.[28] In Arizona, California, Illinois, Iowa, North Carolina, Wisconsin and other farming states, citizens have brought suit against CAFO hazards including "stench from waste lagoons, sinking property values, tainted groundwater and swarms of flies." But lawmakers in those states have passed farm-friendly laws to make it difficult to sue, and the federal government seems to have abandoned its obligation to step in. The citizens' suits thus employ "new liability theories . . . to counterbalance what environmentalists and plaintiffs' lawyers say are often tough court battles and a void in national and local regulatory enforcement." This must be done since, of the 19,000 CAFOs in the US, perhaps three-quarters discharge waste into waterways. It has been hard to prove standing in some cases, but there have been a few successes under the mounting piles of manure.[29]

Sweet Home, Alabama

A major broiler-producing state, Alabama, raises 1 billion chickens annually, selling internationally, including leg quarters to Cuba and chicken feet to Asia, and producing a half-pound of shit for each pound of flesh. Alabama may get employment from the chicken farms, but its citizens get responsibility for dealing with the waste, while the corporate headquarters of virtually all of the money-making big poultry companies sit in other states. Alabama gets the shit, more than 1 million tons annually, but the companies own the birds. And, until 1999 – 29 years after the passage of the National Environmental Policy Act (NEPA, 1970) – no regulations specified how to deal with CAFO waste. Each farm adopted its own practices. Even if many farmers were diligent in trying to stem the incessant flow, most did not. Even 15 years later, no progress had been made. Excessive concentration of N, P and other nutrients increasingly contaminated streams, rivers and groundwater, leading to ecological degradation, posing risks to fish and other aquatic life and to human health – through, for example, nitrate contamination in drinking water.[30]

10 *The removal of poultry manure from an industrial shed.*

11 *The major way to dispose of all the chicken shit is by spreading it on adjacent fields, done here by a manure spreader on a French farm. This method is running out of adjacent fields.*

Alabama's chickens, absentee landlord corporations, its topography and waterways, and some 83,000 acres of lakes and ponds were the perfect storm for CAFOphilia that came with mild winters, annual rainfall of 142 cm, and significant runoff that was treated as "nonpoint" pollution.[31] One such Alabama farm, Gilley's, raises chickens (and cattle) for Tyson on their farm. Gilley and wife Terri must deal with the manure produced from up to 125,000 chickens housed in six 40 x 500 ft sheds. In the 14-day interval between dispatching chickens and the arrival of the next load of chicks, they clean the sheds of the wetter waste ("cake"). The drier waste, largely straw and shit, may be removed once a year. The cake will be dried in stack houses, along with dead fowl that have been scooped up, too, and dry waste. Once the mass is dry enough to be moved, it will be transferred to storage for later use or spread on land. Here it begins its seeping into the Alabama ecosystem as runoff powered by frequent heavy rains.[32] The state requires buffers between the manure, farm, waterways, neighbors and roads. CAFOs must file annual inspection reports, including soil-test results provided by private inspectors certified by the state. But the problem is unabating.

In 1999, Alabama finally introduced oversight and regulations to the factory farms. But there have been few results. Fifteen years later, factory-farm agriculture had impaired 515 miles of the state's rivers and streams, with chickens producing the greatest amount of manure, largely in farms in the north of the state. Aerial photographs show the concentration of nutrient waste along waterways; the waste is spreading and flowing. Agriculturalists have "encouraged spreading out the chicken waste, trucking litter away from areas of 'excess production to areas of greatest need.'" Poultry "litter" has been applied heavily and year-round to cropland and pastures adjacent to poultry farms for years. The question in Alabama – as anywhere else – is how to arrange the problem of the proper disposal and management of billions of tons of litter – in Alabama's case, over 1.6 million tons of litter generated at nearly 4,000 poultry farms annually? The transport, drying, spreading hardly make a dent. "Overapplication" is the rule, followed by runoff and pollution – another rule.[33] Yet jobs and business hold sway over environment and public health in Alabama.

According to factory-farm spokespeople, contractors can find another income stream in shit. It's a stream of something, to be sure, but not income. They can sell the manure to other farmers. Yet few Alabama poultry farmers have managed to turn shit into dollars. In 2002, Alabama introduced a ban on the spreading of animal manure on the state's pastures and cropland from Nov. 15 to Feb. 15. Certified animal-waste vendors, who clean out poultry houses and spread the manure as fertilizer on pasture and cropland, worry that this will force them out of business.[34] Some fecal-matter specialists argue that winter application is a viable practice. Yet, clearly, more energetic responses are required than simply buying and selling shit, even if it is good shit. Something must be done with the growing numbers of farm animals in CAFOs of all kinds, and the destruction of groundwater and surface water by manure. Listen – 1 hen produces 130 pounds of manure in 1 year, and 1,000 hens will produce 65 tons.

Economic analysis confirms the environmental costs of CAFOs – and the fact that many of them are passed on to workers and consumers. Union of Concerned Scientists specialists documented how geographic concentration of CAFOs, overuse of chemicals, and excessive size and density exacerbate problems of air and water pollution. If CAFOs are 5% of all meat-animal operations, then they now produce 50% of our meat and 65% of the shit, some 300 million tons/year. The analysis revealed that medium-sized operations were nearly as cost-effective as large ones, not even taking into account the untold costs of pollution and the lack of a social safety structure for CAFO operations. But the tendency is to privilege the large ones with cheap inputs of water, energy and feed (feed being 60% of the total costs) that offset high capital costs, especially since these are passed on to the contractors.[35]

The challenges have grown so great in the face of the growing piles of shit that specialists have suggested dumping the stuff in the forest – on forest soils. A study conducted at NC State Extension Service, whose authors included a Weyerhaeuser Company employee (Weyerhaeuser is one of the largest private owners of timberland in the world), explored such a possibility and concluded with enthusiasm. In the same way they believed they could perfect the meat chicken, they believed they could improve the forest by combining the two. First, they noted that most forest soils in the southern United States are almost universally deficient in nitrogen (N) and phosphorus (P). Then they located the majority of these forests near animal feeding operations that produce large surpluses of N and P in animal manure, noting that North Carolina has 30 counties "where more P is produced in animal manures than can be assimilated by all the cropland and pastureland in those counties, and 7 counties where more N is produced than can be assimilated." They concluded that a win-win alliance between the chicken and the pine tree – or at least their owners – existed, which could be brought together by the manure. Given that North Carolina has around 7 million acres of forestland (58 percent of the state's total land area), the potential for disposal is great. Given nearly 3,000 poultry farms are located in North Carolina producing turkeys, chicken broilers and eggs, that's a lot of shit. How much shit? The state's 760 million broilers and 35.5 million turkeys produce 8 to 10 million tons per year of manure. Finally, the authors note that negative impacts on streams, wetlands and riparian buffers must be avoided when applying animal manures, and appropriate application setbacks must be observed. At a minimum, no waste material should be applied in wetlands and in poorly drained or organic soils; within 100 feet of a well; within 200 feet of a dwelling other than those owned by the producer; within 75 feet of a residential property boundary – that would be great for the neighbors; within 75 feet of an intermittent, seasonal or perennial stream or river, other than an irrigation ditch or canal.[36] These all seem both minimum requirements and difficult to achieve.

Birds of a Feather

What of dead birds and their feathers? Billions of birds die in production each year, many individually, some pecked to death, others from skeletal deformities, others to starvation, and many because of bacterial outbreaks, with the vast majority of mortalities in the industrial slaughter process. Industry representatives are cognizant of these losses and try to limit them, of course. But even small losses result in a massive waste problem. CAST (the not-for-profit Council for Agricultural Science and Technology) notes that a flock of 50,000 broilers grown to 49 days of age and averaging 0.1 percent daily mortality (4.9 percent total mortality) will produce approximately 2.18 tons (metric) of carcasses. Hence, with billions of birds produced annually, producers face great problems in disposal simply because of the vast amount of organic material ("carcasses") that must be handled without threat to the environment or human health.

One question is how best to handle the carcasses – or, as scientists call it, "mortality management." A carcass without application is a loss of time and money. Thus, scientists have studied how to preserve carcasses for uses as feed, fertilizers and other applications. This is rendering, the transformation of dead birds into valuable proteins, minerals and oils – not for human consumption. The effort to find uses leads to the horror-film scenarios of "acid or base preservatic acid fermentation, and yeast fermentation – may be used for safe and realistic on-farm storage," but that may contribute to "a dramatic decrease in the level of pathogenic microorganisms."[37] The use of "aqueous alkaline hydroxide solutions . . . as an alternative method of mortality management" may enable the "conversion of the preserved carcasses and solutions into an acceptable poultry by-product meal." This can simultaneously hydrolyze feathers and preserve the carcass to prevent further "visible feather degradation and carcass solubilization," and odor production, and inhibit microbial growth.[38]

There are so many carcasses that it is nearly impossible to render or dispose of them safely in these and other ways. CAST offers guidelines for dealing with the lifeless cluckless cluckers. The enormous amount of material that must be discarded reveals just how much energy goes into producing meat, disposing of the offal and dealing with dead ducks and chickens. In 2007, US industry produced 8.90 billion broilers with an average live weight of 2.51 kilograms (kg). CAST analysts, assuming average mortality losses of 5 percent for broilers, determined 468.4 million broiler fatalities. Assuming average weight is approximately half of total end weight, then annual mortality weight is 587 million kg or 587,000 tons (1.296 billion lb) for broilers, and this does not account for catastrophic losses that could be encountered during periods of disease outbreaks, natural disasters and so on,[39] like the millions of birds who drown in hurricanes and floods, or that must be culled and properly disposed because of Avian flu.

Whatever the tonnage, this is a massive amount of organic matter "that requires environmentally and biologically safe disposal or use during the course of a normal production cycle." Industry routinely disposes of carcasses through burial, incineration, composting and rendering. But burial is not permitted everywhere because it likely will have negative impact on groundwater. Incineration is "a biologically safe method, but it tends to be slow and expensive and may create air quality issues." Composting has become more popular, while "removal of poultry carcasses from the farm and subsequent transport to a rendering facility offers great potential, but the spread of pathogenic microorganisms during transport is a significant concern." Given global warming and increasingly volatile weather, we should expect to see more bird carcasses encountered from flooding, disease outbreaks and so on.[40] The do-nothing US Congress does nothing on climate, without exception siding with industry, and as a result farmers are stuck with few reasonable solutions to deal with carcasses. How many such carcasses have begun to overwhelm ecosystems across the globe?

And then there are feathers – but broiler feathers, not goose feathers that might be used in down jackets or comforters. The European Union produces 13.1 million tons of poultry meat – and currently 3.1

million tons of feather waste annually, an amount expected to increase as poultry consumption increases. Most feather waste is disposed of in landfills, some incinerated, and only a minor part converted into low-nutritional-value animal food. A research program, Karma2020, therefore sought to develop industrial conversion methods for turning poultry feathers into such valuable raw material as hydrolyzed keratin, bioplastics, flame-retardant coatings, spun bonded non-wovens and thermoset biobased resins.[41]

The quantities of shit, feathers and other organic material contribute mightily to the rapid spread of such bacteria as *Campylobacter*. Once again, rather than take this persistent and growing problem of feathers that may spread dangerous bacteria as evidence of the need radically to change factory-farm production practices, researchers instead have sought modest modifications to industrial practices. In one test, scientists set out to determine whether, during transport and holding of broilers prior to slaughter, conventional solid or elevated wire-mesh flooring influenced the number of bacteria from both feathered and defeathered carcasses. As is usual practice, they took 7-week-old broilers off feed for four hours before slaughter (so there would be less fecal matter in the gut to spread bacteria during slaughter). These were transported at "commercial density" in a modified transport dump-coop, with one side having fiberglass sheeting and the other 2.54 × 2.54 cm (one-inch square) wire mesh flooring that allowed feces to drop. The broilers were transported for 1 hour, held for 13 hours in a shed, electrocuted, and then slaughtered and sliced up – with the vents plugged to prevent escape of feces. They did this Kentucky fried experiment, obtaining carcass rinses both with birds of a feather, scalding and dying together, and after defeathering, with removal of the head and feet. Of course, there were greater numbers of total aerobes, coliforms and *E. coli* on the feathered than defeathered carcasses for both floors. The *Campylobacter* count was also less for defeathered than feathered carcasses from the solid flooring treatment, but that count did not significantly decrease following defeathering of carcasses from the wire flooring. The percentage of *Salmonella*-positive carcasses remained constant. That is, the broilers transported and held on solid flooring had noticeably dirtier breast feathers and higher coliform and *E. coli* counts prior to scalding and defeathering, but bacteria recovery from external carcass rinses did not differ between the solid and wire flooring treatments after defeathering.[42] One could prevent these problems in the first place by cleaning facilities more frequently, packing them less tightly, and avoiding significant bacterial infestation in the first place.

Broilers Warm the Atmosphere

This chapter has attempted to document the high costs of factory farms from an environmental perspective. I have not thoroughly examined each and every nuance, but it is now clear that unresolved issues in disposing of waste have had a significant impact on many communities – and nations – and they are being passed on to consumers or ignored. Livestock, of course, contribute significantly to climate change with the billions of metric tons contributed to greenhouse gases – perhaps 18 percent annually worldwide according to some sources, and, more recently, perhaps as much as half of the emissions,[43] suggesting that chicken shit, and other forms of shit, and methane burps and farts, are indeed dangerous to living things. There is no doubt about this. Livestock contribute significantly to climate change through emissions, to deforestation, to desertification and so on.[44] A brief examination of the contribution of factory farms to this growing crisis follows. I consider beef-cattle and other meat-animal farming in addition to the broiler industry.

It is well known that beef cattle raising is the most unsustainable agricultural practice in the world. Livestock production accounts for the majority of global agricultural land use and contributes around 14.5 percent (one-seventh) of global greenhouse gas emissions. In Brazil, cattle ranching is responsible for half the country's emissions, and 80 percent of deforestation was associated with demand for animal pasture between 1990 and 2005. In 2019, devastating forest fires connected with illegal deforestation practices in pursuit of land to graze more cattle have brought additional environmental havoc to Amazonia. Moreover, meat and dairy production uses around 8 percent of the total water that humans use, much of it toward cultivation of such feed crops as soy. Brazil's agriculture production and pollution problems exist in other major producing countries, no matter what the efforts to regulate pollution of land, water and air, prevent deforestation, and make

CAFOs more humane and environmentally sound. According to the US Environmental Protection Agency, agriculture has polluted more than 230,000 km of rivers and streams, and some 400,000 hectares of lakes, bays and more. And the future looks more gassy, grassy and less forested as demand for livestock from China's growing middle class will increase consumption another 70 percent by 2050. And all of this goes hand in hand with such intensive farming methods as CAFOs, that, as discussed here, involve dubious safety, sanitation and animal health quality, and are highly polluting.[45]

Long-term growth in the international trade in meat since 1990 means that methane (CH4) and nitrous oxide (N_2O) emissions from beef, pork and chicken produced in one country are increasingly related to meat consumption in a different country. These large transfers of emissions occur between North American countries, from South America to Russia, and between European countries. An evaluation of non-CO_2 emissions from beef, pork and chicken produced in 237 countries reveals that, over the period 1990–2010, an average of 32.8 MT (metric tons) CO_2-eq (carbon dioxide equivalent) emissions were embodied in beef, pork and chicken traded internationally; the quantity of these emissions increased by 19 percent in this period. The largest trade flows of emissions embodied in meat were from Brazil and Argentina to Russia (2.8 and 1.4 MT of CO_2-eq, respectively). Trade flows within the European region were also substantial: beef and pork exported from France embodied 3.3 MT and 0.4 MT of CO_2-eq, respectively. This all supports the contention that trade may result in an overall increase of greenhouse gas emissions when meat-consuming countries import meat from countries with a greater emissions intensity of meat production, rather than producing the meat domestically.[46] On top of this, the amount of global non-CO_2 emissions embodied in chicken rose faster than those of beef and pork, and were a substantial percentage of those total emissions, with the US and Brazil, and to a lesser extent the Netherlands, the major contributors to these emissions.[47] As the authors of a paper on this phenomenon pointed out, international food trade can play a role in enabling local and regional food security, and meat trade may be environmentally beneficial when it flows from resource-abundant countries to resource-scarce countries. Yet there are such risks involved as well, including the growing dependence on exporting countries (such as Brazil, France, Argentina) and precisely such significant and growing environmental impacts as emissions of greenhouse gases.[48]

And there are still other environmental costs that come from direct and indirect subsidies to meat-animal industries. According to Gurian-Sherman:

> Properly managed pastures, for example, require less maintenance and energy than the feed crops (such as corn and soybeans) on which CAFOs rely. Healthy pastures are also less susceptible to erosion, can capture more heat-trapping carbon dioxide than feed crops, and absorb more of the nutrients applied to them, thereby contributing less to water pollution. Furthermore, the manure deposited by animals onto pasture produces about six to nine times less volatilized ammonia – an important air pollutant – than surface-applied manure from CAFOs.[49]

The problem is that feed grain subsidies from federal farm bills let grain prices fall below the cost of production and compensate farmers for the difference. Thus, taxpayers (the government) pay, providing indirect subsidy. Gurian-Sherman indicates the huge costs in the US: "When extended to include the dairy, beef, and egg sectors, low-cost grain was worth a total of almost $35 billion to CAFOs from 1996 to 2005, or almost $4 billion per year."[50] In essence, CAFOs can rely on a holdover law from the Depression when farmers needed support, so called Title-I payments to commodity-crop farmers that compensated them when the price of corn, soybeans or some other commodities fell below their production costs. Without these subsidies, commodity-crop farmers would not be able to stay in business for long. CAFO operators directly benefit from the situation.

Another measure of the true costs of CAFOs concerns calculations for pollution losses and property value losses. As discussed, the manure problem leads to runoff and leaching of waste into surface and groundwater, "contaminating drinking water in many rural areas, and the volatilization of ammonia. Several manure lagoons have also experienced catastrophic failures, sending tens of millions of gallons of raw manure into streams and estuaries and killing millions of fish." Smaller but more numerous spills also frequently occur. In Kansas alone, remediation of leaching at dairy and hog CAFOs has cost $56 million to taxpayers. On the basis of this and other data, one researcher calculated the total cost of cleaning the soil under US hog and dairy CAFOs at $4.1 billion. The loss of drinking water and fisheries has been substantial. According to the USDA, it would cost between $134 million and $153 million for the Delmarva region alone to transport manure to enough crop fields and pastures to comply with Clean Water Act rules – or perhaps between 43 and 49 percent of net returns.[51] Another study determined that each CAFO in Missouri has lowered property values in its surrounding communities by an average total of $2.68 million. Extrapolating by the total of the 9,900 CAFOs in the US, property losses might total $26 billion.[52]

What can be done about CAFOs, about factory farms, about vast quantities of manure, offal and feathers produced in the manufacture of billions of broilers? This chapter has considered only the extent of the problem and does not pretend at solutions. It notes that, to date, all measures considered have barely had an impact in reducing pollution and managing waste. In fact, the pollution problems are growing, and the public health and safety problems connected with them grow, too. We see this in our waterways, our air, and in our food. CAFOs have been able to pass costs along to nearby citizens and to far-away citizens. They have taken advantage of inadequate legal statutes and regulation. They have relied on the low cost of products in the supermarket to deflect attention from exogenous costs. And wherever CAFOs and factory farms exist, they have significant local, regional and international environmental impacts because that is the nature of this beast. A recent Pew study called for caps on total animal density; shared financial and legal responsibility for proper waste management between farmers and corporate integrators; closer monitoring and regulation of waste transported from CAFOs; and a requirement that all medium and large CAFOs obtain Clean Water Act permits.[53]

Animals and waste overwhelm the land with no clear answer how – sustainably – to store and dispose of manure and animal waste. No such thing as natural cleaning or dilution could occur. Could it be rocketed into space, as some hubristic physicists suggested in the 1950s as a solution to radioactive waste? Could contractors potty-train the meat animals? We feed cattle chicken shit, often in combination with feathers and uneaten chicken feed, as a way to reduce the vast quantities of waste and lower feed costs. But a typical sample of poultry litter might also contain antibiotics, heavy metals, disease-causing bacteria, and even bits of dead rodents.[54] So this is not an option, nor is it possible to raise enough cattle to eat all of the manure. The amounts and challenges are enormous – in 2007, some 500 million tons of manure was produced in livestock and chicken CAFOs, perhaps some 3,000 pounds per person. Humans produce far less waste, and at least theirs is highly regulated.

In CAFOs, animals do not deposit their manure directly on fields – let alone poultry potties – nor has waste disposal been regulated according to rates of application, season, or requirements for the reporting, analysis or monitoring of applied manures. Instead, as the number and size of operations has increased, so too has the urgency to act forcefully to regulate and manage the shit problem. This is no longer a byproduct, but a significant hazardous waste problem, and federal and state regulators and environmental officials have not figured out a way to manage it.[55] Nor, frankly, have CAFO owners seen any reason to hurry along in finding a solution. And who knows how expensive it will be to remediate pollution and impose restraint on shit and other waste across the globe as billions upon billions of animals are shoveled out of factory farms along with their shit each year. And just remember that there are allowable levels of fecal matter in your food. No shit.

12 *Broilers trucked in cages to slaughter. From hatching to death, they are boxed and caged to prevent motion – except as impelled along by machine.*

5

Pecking and Protest

> Now comrades what is the nature of this life of ours? Let us face it: our lives are miserable, laborious, and short. We are born, we are given just as much food as will keep the breath in our bodies and those who are capable of it are forced to work to the last atom of our strength: and the very instant that our usefulness has come to an end we are slaughtered with hideous cruelty.
>
> – George Orwell, *Animal Farm*

Pushed and prodded with words, not electrical stun guns, the European governments have sought to use democratic means to ensure the better welfare of meat livestock. Russia recently passed a law seemingly intended to enforce proper treatment of domestic animals that may have a small influence on improving factory-farm practices. In the US, pushback against the forced march of production in factory farms has begun to resonate beyond sheds and batteries into the public sphere and in restaurants. In a variety of ways, the plight of the broiler chicken has given rise to protest about its treatment. This protest, broadly defined, extends from the early days of factory inspections to protect public health over fears of diseases and adulterated products into the late twentieth century when animal welfare became a crucial issue in factory farms. If early concerns about the plight of factory-produced meat animals concerned foodborne diseases, then they now concern the very system in which fowl are considered machines, not animals. Animal welfare has become a statutory issue that mediates the relationship between farms, producers, sellers, products and consumers. This chapter offers commentary on the cacophony of sounds surrounding 21st-century meat animals, including efforts to raise them in more humane fashion.

In some places, animal welfare considerations are not only on the back burner, but considered a nuisance. In the United States, for example, egged on by legislators in a number of agricultural states, industrial interests have sought to quiet the growing din of animals confined in factory farms in the name of profits by seeking to make it illegal to share videos and reports about inhumane production practices. In a series of such reports, whistleblowers and undercover workers have recorded mistreated animals, howling in pain, wasting away, dying, without veterinary care, being kicked and struck by abusive employees who, themselves, are forced to become inhuman to meet production targets.[1] How striking, therefore, that producers have decided to block public scrutiny of their absent animal welfare practices, so that the consumer sees only the finished, quiet, packaged product in the supermarket display case. The producers claim that the reports are fraudulent or produced by disgruntled employees, and thus that the chick's peep and the pig's squeal should not be revealed to the public in the name of preventing "agricultural fraud." The philosopher might ask: If those sounds cannot be heard, do they still not echo across the feed lots and slaughterhouses in the suffering of animals? The ethicist might chime in that, to treat animals as objects, we undoubtedly treat ourselves as lesser creatures, devoid of humanity, without concern for the suffering of others, and potentially also creating hierarchies of worth between and among species, including ourselves.

The effort to silence those who would deign to describe the houses of horror and sheds of shit dates to the late nineteenth century when many public and private individuals insisted that the modern state use its regulatory powers for the general good and to employ science and engineering toward the ends of conservationism and public health. Increasingly anchored in the germ theory of disease, state commissions determined the need for state-supported supply of clean water, the pick-up of rubbish, the inspection of medicines and foods, and so on. In

Massachusetts, officials, physicians and other scientists came together finally to develop plans to provide clean water to the burgeoning metropolis of Boston on the Atlantic Coast, building a water transfer network of aqueducts and reservoirs to provide fresh water from New Hampshire and Western Massachusetts. In Prussia in the 1880s the state introduced health insurance, accident insurance in 1884, and retirement insurance, while waterworks also spread quickly, although perhaps in response to rising incomes, not a public health crisis.

Yet, in spite of these important steps forward, industries continued to have the upper hand in avoiding regulation of their commerce until press muckrakers opened the doors of their maiming factories and polluting effluent pipes to public scrutiny. One such muckraker, the journalist Upton Sinclair, penned a novel, *The Jungle* (1906), about the dangerous and dark lives of immigrants to Chicago, many of whom worked in dreadful conditions in the meatpacking industry. Although fictional, Sinclair's exposé of life in the industrial abattoirs of Chicago uncovered the hell of the food industry. The unhealthy, dangerous and unsanitary practices led to the Meat Inspection Act of 1906 which prohibited the sale of adulterated or misbranded livestock, and improved the conditions in factories. The 1906 Pure Food and Drug Act that prohibited the manufacture, transport and sale of adulterated food, alcohol and drugs followed. Like today's animal rights activists, Sinclair worked incognito in meat-packing plants and stockyards to research his book and bring to light animal immorality and worker exploitation, publishing the novel first in serial form in *Appeal to Reason* (a socialist newspaper) in 1904, with Doubleday publishing the book in 1906.

The reader of *The Jungle* learned of the hellish side of food and labor through the experiences of Jurgis Rudkus, a Lithuanian immigrant, and his wife, Ona, whose attempts to start a new life in America fall prey to injury, poverty, exploitation, the swindles of conmen, rape and death, and imprisonment. Rudkus discovered factories not as liberating, but as killing machines, each as hell on earth, and, unlike the heat- and humidity-controlled chicken sheds of the twenty-first century, Chicago abattoirs were places of blistering cold and burning heat, a place where a man might fall unnoticed into a boiling vat and be turned into canned food. The workers who survived lost limbs and faith, and when they had been entirely sapped of energy, too, were discarded like the animals.

Like today's CAFOs, Rudkus's meat-packing plant was not a place fit for man or bird.[2]

It may be that Rudkus's Lithuania, building on the foundation of the dilapidated Soviet chicken industry, has now found a better path. The company Dovainoniu Paukstynas, established after the fall of communism in 1993, having settled upon hatching and selling 1-day-old Cobb-500 broilers for their growth and succulence characteristics, and working closely with Agrotech (Holland) to provide modern feeding, watering, heating and ventilation systems, and with farms in Staniunai, Prazariskes, Kairiskes, Lapainia, Morkunai and Paparciai, Lithuania, will avoid the human costs of meat production that Rudkus barely survived. Indeed, another Lithuanian company with operations in Vilnius and Kaisiadorys that hatches over 48 million broilers annually for human markets, claims that broilers are raised strictly to EU welfare, hormone and antibiotic standards, and "sometimes poultry experts joke that they seems to care more about chickens than humans."[3] There is a need for greater scrutiny for CAFOs precisely because of work conditions, animal welfare conditions and disease vectors that continue to press upon human and animal health into the twenty-first century.

Chicken welfare officers, often zoologists, recognized the need for measures to lessen the suffering of their feathered wards long ago, although it was only in the 1990s that they began to push from within the industry to establish norms of behavior and industrial practices to reach that goal. In a way, these individuals were assisted in the goal by animal activists and NGOs who sought, through publicity, to raise the specter of the cruel treatment of these animals, at times through undercover investigations by the activists who took on work at poultry farms, and by film and other means documented the cruelty, filthy conditions, diseased animals, and mistreatment of workers who had bought into the industrialization of fowl and considered birds not fowl, but output.

Avian Protest

Consumers have long reacted with horror to food and drug scandals, although one wonders whether they have become inured to safety concerns in the twenty-first century where chickens move like clockwork from egg to cellophane. To be sure, many consumers around the globe have thrown their hats and stomachs into the circle with fast food. They gobble up high-fructose corn syrup, and they are salted, sugared and baked with trans-fats in a variety of foods, in many countries becoming increasingly obese, or suffering from high blood pressure, or having to deal with type 2 diabetes and so on. Yet there have been several pauses at the industrial dinner table to take in an aperitif of health warning since the 1990s when they have learned about the costs associated with monocultures, factory farms and other new technological phenomena, and as they worry more and more about potential risks of GMOs and antimicrobial resistance.

It may be that the "Chernobyl event" in food awareness was the Bovine Spongiform Encephalopathy (BSE) – or Mad Cow Disease – scandal. It has spilled over into the arena of pigs and broilers, especially since the rise of international pandemics of Avian Influenza. The BSE crisis of the 1980s and 1990s, born through the spread of prions into the meat of cattle after they had eaten the spinal cords and other nervous tissue of hundreds of thousands of infected animals (cows) that had to be culled. It has killed somewhat over 200 people worldwide, but the horrific nature of the disease – dementia, coma, death, without cure – shocked even employees of the food industry intent on feeding spines to ruminants.[4] Factory farming was a direct cause of the BSE scandal, in its effort to harness every morsel of shit or dead animal to the food chain. We have arrived at a paradox where factory farmers insist we must regiment the animals' lives – hold them, push them, monitor them, feed them, but deny them normal habitat – precisely to prevent risk to public health. And yet, in denying them welfare, we seem to make our food supply more difficult to manage safely.

One of the ways to improve animal welfare and to ensure safer and better food is to lobby and pressure food producers, from the CAFOs to the restaurants, to adopt humane procedures. UK World Animal Protection (WAP) evaluated animal welfare practices in the way fast-food companies protect and manage their chickens at a number of major franchises: Burger King, Domino's, Starbucks, KFC, McDonald's, Subway, Pizza Hut and Nando's. In a worrying, brief study of the fast-food industry, UK WAP considered and compared the massive scale and volume of production, the popularity of the companies, their products, and other factors to determine that none of them had an "effectively-implemented strategic commitment to chicken welfare," although some of them had started to develop chicken welfare policies. UK WAP worried about the absence of "tangible objectives and targets" to deal with this else, considering how important welfare was for some 50 billion chickens harvested annually from factory farms. The chickens were newly engineered, since 1957 experiencing a fourfold increase in kill weight, too quickly, and in overcrowded misery – crushed together in unnatural, boring environments, of artificial light and ammonia-laden litter in which they scratch and bathe. At best, according to UK WAP, the restaurant rankings ranged from poor and very poor to failing, for all of the restaurants.[5]

Debeaking of Public Discourse

Worldwide, there may be some 180,000 NGOs or groups, large and small, that are concerned with animal welfare. Some are international, others national and local. The groups seem to be most active in Europe – perhaps Britain's activists are the most deeply engaged; I learned a great deal from the work of Compassion in World Farming.[6] They all do important work. Here I want to examine briefly the glacially slow industry response to growing pressures to consider animal welfare in their production practices. I am interested in particular in "ag-gag" laws in the United States as a way to stifle speech about CAFO safety. "Gagging" efforts to expose illegal or immoral practices in animal husbandry is the debeaking of public discourse. These laws reveal the effort, strictly on behalf of large corporations, to prevent consumer knowledge about the production and delivery of the foods they eat, without themselves eating into profit margins from producing each unit of flesh. They consist of efforts to stop reporting on CAFO practices, and likely violate free-speech laws. Squawk!

Things are bad for an industry when its representatives lobby ("payoff") government officials to pass legislation to make it illegal to report on health and safety problems in that industry. In the United States, this attempt to put blinders on the public – no less than vision-distorting lenses on chickens – sought to hide the smells, sounds, squawks and poisons of the increasingly violent, dangerous and environmentally risky nature of concentrated animal feed operations. Measuring every input, pushing the animals to fatten as quickly as possible, packing them tightly to keep costs down, not permitting them to move about to save space and train all energy into fattening, the designers of CAFOs have created sheds of torture hundreds of square meters in area that have spread across the earth's surface to meet growing consumer demand for meat. As we have noted, the carcasses, offal, urine and manure are jettisoned to the side in lagoons or other permanent and temporary storage facilities prone to leakage, and without a doubt unsafe and unhealthful for people and the environment. Should the public not know about whence their meat?

If the effort to criminalize publicity of animal cruelty and food safety was in the public interest, then it would occur in places and times where the public is defined narrowly – in the provinces of Robber Barons, dictators and the nineteenth century. Yet, in a series of places, primarily in the land of democracy – the United States – industries, working closely with representatives in agricultural states, have tried to stifle public discussion of animal welfare by precluding investigation of it in the first place through ag-gag laws.

Whistleblowers have exposed factory farms by using video recordings and other means to publicize the farms' practices. These exposés have resulted in negative publicity for the companies and have been bad for factory-farm business. In an effort to combat attempts to expose their practices, not to mention to obscure the violent nature of chicken and other kinds of farming, businesses and friendly legislators have joined to silence them. Ag-gag laws in the US date to 1990 when Kansas, Montana and North Dakota passed laws to protect large-scale agricultural operators. After 2010, with legislatures and governorships in Republican hands, about half the state legislatures in the country considered ag-gag laws, and these were passed in Iowa, Missouri, Idaho and Utah. In Kansas, legislators passed the "Farm Animal and Field Crop and Research Facilities Protection Act" that goes further than existing trespass and other laws to make it illegal for someone to enter a private enterprise to take pictures or video, with felony charges of property damage exceeding $25,000. Building on the momentum of court decisions to reject other such more recent state laws, the Animal Legal Defense Fund, joined by the California-based Center for Food

Safety and Kansas-based farmed animal organizations, sued Kansas to challenge the constitutionality of the law in December 2017. Lawyers successfully argued that ag-gag laws "criminalize innocent behavior and interfere with the free speech rights of reporters and whistleblowers."[7]

Some of the legislation appears to be inspired by the American Legislative Exchange Council (ALEC), a business advocacy group with hundreds of state representatives from farm states as members. The group creates model bills drafted by lobbyists and lawmakers, which in the past have included such things as "stand your ground" gun laws and tighter voter identification rules. ALEC is intended, according to its website, to give a voice to corporations and their interests; most of its funding comes from corporations. One of the group's model bills, "The Animal and Ecological Terrorism Act," prohibits filming or taking pictures on livestock farms to "defame the facility or its owner." Violators would be placed on a "terrorist registry."[8] This model legislation seems to have been the foundation for ag-gag laws passed in Iowa, Utah and Missouri that would make it nearly impossible to produce undercover film or other exposés of illegal animal practices.[9]

One law passed in North Carolina in 2016 enabled companies to sue whistleblowers who recorded or took pictures of "wrongdoing." By punishing whistleblowers, ag-gag laws thereby protected factory-farming companies at the expense of animal welfare and public health. State Rep. John Szoka (R–Fayetteville), the bill's chief sponsor, claimed the bill would not prevent "regular" employees from reporting illegal workplace activities or practices, although it established a civil "right of action" for the owner or person in lawful possession of property if that property is wrongfully taken or carried away. In other words, it enabled the civil prosecution of anyone who entered the "non-public area of another's premises, including an employee who captures or removes data, paper records or documents and then uses the information to breach the person's 'loyalty to the employer'" – or someone who recorded images or sounds in non-public areas.

How far meat producers have come since 1906 when Rudkus nearly died in a meat-packing plant! Now the chickens' enemy was the North Carolina General Assembly, that passed a law (HB 405) to allow whistleblowers to be sued for secretly taking pictures in the workplace or exposing what they define as "trade secrets." These might be employees trying to document safety concerns or animal activists engaged in revealing animal abuse. But producers striving to keep costs down – and the manure, offal and screams from the CAFOs from getting into public view – have convinced legislators to prevent public disclosure. In this way, they can pay laborers less, push them harder, turn them into machines themselves, and quickly, if painlessly, bleed, slice and quarter living creatures into food items.

Surprisingly, Governor Pat McCrory (R), a man who would sign almost any bill the Republican majority sent to him, vetoed the bill. Was it McCrory's conscience? His worry about animals? After all, he signed bill after bill to deny voters' rights, and he worked tirelessly to see the so-called Bathroom Bill passed – a bill to discriminate against transgendered individuals by requiring people to go to bathrooms for the sex listed on their birth certificate.[10] But chickens, like gays, lesbians and transgendered individuals, were treated as subhuman and were out of luck. McCrory's Republican colleagues overrode his veto, on June 3, 2015. The expansive bill was directed only toward protecting the factory-farm industry, but, in an effort to obfuscate its intent, it covered all North Carolina businesses where public scrutiny of wrongdoing was in the public interest, including nursing homes, day care centers and so on. In this way, the legislature put the elderly, children and chickens in the same category, with the clear intention of punishing whistleblowing on wrongdoing perpetrated against these defenseless creatures.

As in the day of Upton Sinclair, so in the twenty-first century activists sought to expose this wrongmindedness and sued in the Federal courts to overturn HB 405 as a violation of the constitutional protections of free speech. They have won in every case, although at some financial cost, and in some states the legislatures, chained like pigs to cages, have appealed the decisions. In North Carolina, PETA, Mercy for Animals, the ACLU (the American Civil Liberties Union) and others brought suit. After several setbacks, in June 2018 a three-judge federal appeals court reinstated the lawsuit, determining that a violation of the constitution had occurred since the state, by threat of its direct action, stopped groups "from conducting undercover investigations and making public statements."[11] There is precedent to support the plaintiffs, namely the decision of a federal judge in August 2015 that struck down an ag-gag law in Idaho on free-speech grounds, the first such ruling in the country. Activists who pose as employees to gain access to farming operations, the judge wrote, "actually advance core First Amendment values by exposing misconduct to the public eye and facilitating dialogue on issues of considerable public interest."[12] Fortunately, to date, all ag-gag laws have been declared unconstitutional under the free-speech protections of the First Amendment. Yet anyone who violates an existing law – say, by secretly taping abuses of elderly patients or farm animals and then sharing the recording with the media or an advocacy group – can be sued by business owners for bad publicity and be required to pay fines and face incarceration.

Another way in which the manufacturers of meat have attempted to silence objections to factory practices is through the courts – for example, using SLAPP (strategic lawsuits against public participation) statutes. In a SLAPP action, the plaintiff brings a lawsuit "to censor, intimidate, and silence critics by burdening them with the cost of a legal defense until they abandon their criticism or opposition."[13] Usually, it is a powerful and financially flush organization, corporation or individual who brings such a lawsuit; in a number of jurisdictions, SLAPPs have been made illegal on the grounds that they impede freedom of speech. Tyson Foods, for example, faced suit in California for making, according to the nonprofit PCRM (Physicians Committee for Responsible Medicine), false and deceptive health claims about its chicken products sold in California. Recall that Tyson sells more than 25 percent of the total chicken meat products consumed by Americans, so this is a lot of chicken and a lot of false claims. PCRM alleged that Tyson engaged in two advertising campaigns replete with false and deceptive statements about its products, in violation of the Business and Professions Code. Tyson portrayed chicken meat as a "heart-healthy" food and said consumers could serve chicken "as often as you like" because, its campaign asserted, Tyson's chicken products were low in saturated fats and cholesterol. PCRM claimed that the advertisement created the false impression that chicken "is a health food that can protect against the risk of developing heart disease." PCRM also noted that, actually, most Tyson chicken products contain substantial levels of fat and cholesterol that would likely increase the risk of heart disease. In a second advertising campaign, PCRM pointed out that each ad claimed that Tyson chicken products are "all natural," and were great for children. PCRM argued that, in fact:

> Tyson raises its chickens in a "factory farm" system in which the chickens are genetic mutations that do not exist in nature, the chickens are vaccinated, the chickens are medicated immediately after being hatched, the chickens are crowded together by the tens of thousands under one roof, and the chickens are routinely and regularly fed antibiotics at therapeutic and sub-therapeutic levels to combat and prevent diseases facilitated by the unnatural overcrowding and to stimulate an unnatural growth rate.[14]

The case was decided against Tyson.

In Iowa, such groups as the American Civil Liberties Union of Iowa and Bailing Out Benji – an Ames-based nonprofit that protests against "puppy mills" – have brought suit against Iowa's law prohibiting "agricultural production facility fraud." Like other laws, this one made it an offense punishable by up to one year in jail to obtain access to an agricultural production facility by false pretenses, or to make a false statement in connection with a job application at such a facility. The law duplicated existing trespass statutes, but the goal was to prohibit journalists, watchdog organizations, whistleblowers or others from gaining access to such facilities. Lawmakers had tried to make it a crime to produce or distribute, in any photographs, videos or written materials, descriptions of animal abuse – to prevent undercover investigations of agricultural production. The central notion of this law is that "when it comes to exposing violations and problems in the nation's food chain, the government must protect the private interests of industry even if it means putting the public at risk." The *Des Moines Register* Editorial Board asked:

> These laws impose criminal penalties even in cases where all of the public disclosures are truthful, accurate and in the public interest. In other words, they make it illegal to speak the truth, which is contrary to the fundamental American principle of free speech. How is a livestock operation harmed by the reporting of facts, unless those facts reveal uncomfortable truths?[15]

Ron Birkenholz, a spokesman for the Iowa Pork Producers Association, said the law offers "meaningful protection to farmers from those who seek and obtain farm employment under pretenses," and he argued that the law protected "all citizens' constitutional rights."[16] And in February 2018, the courts allowed challenges to Iowa's ag-gag law to go ahead, for violating the First Amendment. The Legal Director of the Iowa ACLU said that the law had "effectively silenced advocates and ensured that animal cruelty, unsafe food safety practices, environmental hazards and inhumane working conditions go unreported for years."[17]

In state after state, the courts have determined that efforts to ban such free-speech activities as reporting, taping and filming cannot be allowed. Utah's ban on filming meat operations was declared unconstitutional by a US district judge who said "Utah had failed to show the ban was intended to ensure the safety of animals and farm workers from disease or injury." The judge quoted one of the bill's sponsors in the state legislature, Rep. John Mathis, who said the ban was a response to "a trend nationally of some propaganda groups . . . with a stated objective of undoing animal agriculture in the United States," and another, Sen. David Hinkins, who admitted that the bill targeted "vegetarian people that [are] trying to kill the animal industry."[18] What was his beef really about?

Factory farms are hardly without resources. They can always get the assistance of the authorities when in a bind since the chicken is more important than the taxpaying citizen. In winter 1993, the US Army National Guard was ordered by the Governor of Arkansas to fight the evil of 3 inches of ice that blanketed Northwest Arkansas, leaving chicken and turkey farms without electricity. "Operation Save the Chickens" indeed saved the Arkansas poultry industry millions of dollars, as soldiers unloaded portable diesel generators to heat the chicken houses, the soldiers sleeping on cots nearby and often wearing gas masks to tolerate the overpowering smell. Of course, Arkansas has been far less charitable with taxpayers' money when supporting the indigent or raising the quality of education or enforcing environmental protection laws.[19]

Perhaps one of the reasons that animal husbandry states have sought to gag public awareness of the environmental and social costs of confined animal operations, including the horrors of animal cruelty, is precisely the well-educated consumer and resident. Iowa's own pork industry is facing opposition to the construction of more factory-farm facilities in recent years because family farmers are being squeezed out of the work and off their land; industry contraction (fewer, bigger farms, more pigs) means fewer farmers, and thousands of jobs lost in those counties where the shiny, shit-laden production operations arise. In 1977, the USDA counted more than a half-million hog farm operations. Today the latest census reports 63,236, a drop in number of 90 percent. Even if demand for pork and chicken has grown because of rapidly expanding Chinese demand, and that from Japan, Canada, Mexico and South Korea, consolidation continues apace. And the quality of life in the countryside is forever diminished – and has a smell of shit. Said one farmer: "All I ever wanted to do was farm. . . . Then all this shit came up."[20]

The pressure on the massive meat manufacturers has had some impact on inhumane feedlot practices. Smithfield Foods, the largest pork and pig company in the world – with over 500 farms in the United States, plus contracts with another 2,000 independent farms around the country, and facilities in Mexico, Poland, Romania, Germany and the UK, and over 50,000 employees, with one 973,000-square-foot meat-processing plant that processes 32,000 squealing pigs a day, and that sells packaged products under its own name and Farmland, Armour and Cook's – has asked its contracted farmers to end the use of small stalls ("gestation crates" that confine pregnant sows so that their energy goes solely to birthing) by 2022. While not making the change mandatory, Smithfield indicated that it was more likely to renew contracts with those facilities that opt for more humane birthing. By allowing four years for the change to occur, Smithfield can placate pressure from animal rights activists while permitting capital investment in the change over several years. One journalist reported that roughly 10 percent of the country's 5.8 million sows are housed in systems other than a gestation stall; a 2010 study from the University of Minnesota estimated it could cost the pork industry as much as $3.3 billion to make the move from crates to group pens. But the cost will be borne by contractors, not Smithfield, and it must be kept in mind that the vast majority of other CAFOs have yet to recognize the horror of confining a pregnant animal in a cage about its own size so that it cannot move about.

Throughout Europe, protests against industrial agriculture are on the upswing – for the same reasons that activists, concerned citizens and others in the US have struggled to expose the inherently inhumane practices of CAFO agriculture. As in the US, they are worried about animal welfare, pollution, the comings and goings of heavy trucks bringing feed and chicks in and chickens out, but leaving the shit behind, and the extensive social impacts of factory farms on local communities and local ecosystems. In 2011, some 22,000 people demonstrated in Berlin, Germany, to call for changes in the agriculture industry, and especially "for non-toxic and environmentally friendly arable and animal farming, less industrial agriculture and more consumer protection." Such organizations as BUND (Friends of the Earth Germany) worked to bring together people from all over the country, who gathered under the slogan, "We've had enough – no to genetic engineering, factory farming and export dumping."[21]

Another European-based organization, Compassion in World Farming, fights the same battles. It has joined a number of other organizations in the struggle to maintain environmental and social health and stability against the onslaught of the factory farm. If the factory farm must comport with more stringent animal welfare, pollution and public health requirements in the EU than in the US, then still the spread of CAFOs and other factory-farm units has upset local citizens there deeply, and perhaps nowhere else than in France, the leading agricultural power in Europe with a long tradition of support for the family farm.

France supplies 23 percent of the agricultural production in the 15 EU member states as the major producer of cereals, poultry, beef and wine, and with 600,000 farms at an average size of 42 hectares (around 104 acres).[22] Nearly a billion chickens are raised each year in France, over four-fifths of them in factory farms in "dark and bare buildings" where rapid growth and health problems are "daily life." They are crowded in conditions incompatible with their well-being, at 17 to 22 chickens/m^2, a sheet of paper per chicken. Their lives consist of "deterioration," pathologies, poor bone development and so on since they are denied locomotion, and filth. These methods have contributed, Compassion in World Farming France asserts, to serious health consequences for humans, too: to the emergence and spread of Avian flu and food poisoning, problems that will likely grow because of the overuse of antimicrobials.[23]

In 2013, Compassion France joined more than 40 other groups, including farmers' unions, and environmental and animal welfare charities, to protest in Paris to call for an end to intensive farming in France. The demonstration was organized by a local citizens' group, Novissen, which, for its part, has been fighting "a mega dairy project." The protesters' long march ended at the Porte de Versailles exhibition center where the International Agricultural show was taking place, itself a typical propaganda effort intended "to portray a very idyllic picture of farming in France and abroad, and not reflect the reality that around 80% of animals in Europe are raised in intensive systems."[24] Of course, factory farming has a negative impact on climate change; when France hosted the UN Climate Change summit in 2015, farm union activists pointed out that the spread of intensive farming in France had also to be considered as a contributor to climate change precisely because of this impact. In response, the agricultural show advanced the theme of promoting sustainable ecology.[25]

Novissen, which opposes the use of hormones, antibiotics and other chemicals in dairy farming under the banner "Non aux usines à vaches!" ("No to cow factories!") was formed in 2016 to fight a factory dairy farm at Drucat-le-Plessiel and Buigny-Saint-Maclou, near Abbeville, that lives on after the death of its founder and owner, M. Ramery, over concerns about the unhealthful impact of environmental degradation, road accidents, nitrates, pesticides and methane.[26] The protesters sponsored a campaign to "Take a Leak for Glyphosate!" ("Pisseurs involontaires de glyphosate!" or, literally, "Unintentional Pissers of Glyphosate") urging people to be tested for chemicals in their urine, at a cost of €140 per person, to cover cost of the analysis to be made in Germany and the payment of legal fees to file a formal complaint.[27]

The Union of Farmers has been quite outspoken in its criticism of industrial farming in France, discussing its moral and financial costs, and producing a "map of industrialization" that indicates the spread of factory farms across the bucolic French landscape. They criticize the "real intention to deliver our business to the hands of industrialists more concerned with their market share than employment, food or the environment" through the monopolization of farming from the field to production to banking – for example, "the octopus Sofiprotéol,"[28] that "monopolizes production and added value in disregard of the peasants." The union demanded that the government support the farmers and environmental and climatic concerns.[29] They have taken to the streets, blocking Paris with over 1,500 tractors to protest against the destruction of their lifestyle with tax policies, low food prices and "crazy" environmental standards.[30]

Another organization, L214, denounced the conditions of broiler farming in France. L214 members celebrate their vegan practices and accordingly take on animal welfare abuses in factory farming. The association was founded in 2008 for the "Stop Gavage!" campaign for the abolition of foie gras, and has continued its activities in L214. They take their name, L214, from an article of the rural code of 1976 that recognizes an animal as "a sentient being." The goals of L214 include publicizing the nature of food production and conditions of breeding, transport, fishing and slaughter. They "report the reality of the most widespread practices, change them or disappear through information and awareness campaigns. [They] identify and attempt to sanction illegal practices through legal action." Their work is based on the struggle with the concept of "speciesism" – a belief that humans are the superior species and that all others are subservient to their needs – and the demand to cease animal consumption and other practices that harm animals.[31] L214 argue that the intensive breeding of broilers is dangerous for human health. Indeed, they point out, factory farming is a frequent cause of food poisoning, while the profligate use of antimicrobials creates other challenges. They claim that "intensive poultry breeding encourages the emergence of epidemics." They remind the consumer that chicken production is also a source of greenhouse gases, claiming that to produce 1 kg of chicken, 7 kg of CO_2 are emitted, while factory farming also involves profligate use of water: it takes more than 6,000 l of water to produce 1 kg of chicken protein, while vegetable proteins require about 15 to 30 percent less.[32]

In April 2019, L214 published a report – and disturbing video of the appalling daily life of more than 50,000 chickens in an intensive breeding facility in Solignat, in the Puy-de-Dôme.[33] The prefecture of Ain had already received a number of complaints about the farm, leading the Minister of Agriculture, Stéphane Le Folle, to say he would close the facility if need be, given the "lamentable" and "unbearable" conditions. The co-founder of L214, Brigitte Gothière, "fights for a world without consumption of meat and more widely attacks hens in battery" production, especially those sold under the Matines brand (owned by the Avril group connected with Sofiprotéol, which distributes eggs to the major supermarkets in which I have frequently shopped while living in France: Auchan, Carrefour, Casino, Intermarché and Super U), for overcrowded and inhumane conditions. L214 members have carried out protests at one factory at Gaec where they discovered – and filmed – chickens dead or dying, unable to move, covered with flies and larvae, with flies increasing in huge numbers in nearby towns.[34]

Is Protest on Behalf of Broilers Possible in Closed States?

This battle against the industrial will of factory farmers will likely not happen, let alone succeed, in Russia for two reasons: first, the power of state-industrial interests in securing production of oil, gas, metals and other commodities – including agricultural – in the name of national security; and, second, less developed attitudes about welfare. Here, suffice it to say that we can judge the lack of legal, philosophical and other concerns about farm animal welfare in Russia by the grudgingly slow and inadequate creation of statutes to punish animal cruelty in general. Indeed, a law that took effect late in 2018 to much fanfare – and that was touted as a sign of President Putin's enlightened leadership – took nearly six years to pass, and Putin had vetoed one such bill as one of the first steps in his presidency back in 2000.[35] A petition to President Putin at that time, asking him to sign the law to outlaw animal cruelty, noted Putin's disdain for the 2000 statute "On the Protection of Animals against Ill-Treatment." Rather, the petition noted, Putin's Russia had acquired "the features of an all-encompassing barbarous system of mass zoosadism."[36] It noted that "in some cities and settlements, the leaders of the Administrations themselves take an active part in the indiscriminate shooting of animals in front of children and other residents. Small puppies and kittens are drowned in ice holes alive. Bloodied, wounded animals die painfully. Dogs and cats poisoned with the nerve-paralyzing toxins suffer mercilessly until they die."[37] To the petitioners, this all suggested that Russia's joining of the Consultative Assembly of the Council of Europe was a façade.[38]

In dealing with animals, it does not help that the moral authority of the Russian Orthodox Church has lagged behind the times. On October 27, 2007, a priest at the Trinity Lavra of St. Sergius ordered 30 cats to be captured in sacks and taken to a landfill to be bulldozed. The Patriarch of Moscow and All Russia blessed the public slaughter of lambs for Moscow's Tushino balloon festival in 2008.[39] Nor is it good for dogs, cats and broilers that Russian NGOs have come under attack from the authorities. Their members, to be sure, must walk carefully between topics of environmental, public health, and historical concern lest they face charges of consorting with the enemy.[40]

Much of the attention to animal rights in Russia since 2010 has focused on the large number of homeless animals – especially dogs, but also cats, that are abandoned in large numbers. The LAPA organization ("lapa" means "paw" in Russian) that dates to 2013 was founded in the UK, but has a Russian chapter aimed "primarily on neutering of stray and rescued animals and on education of children about responsible pet ownership."[41] The organization came into existence as the nation struggled to deal with the large number of feral cats and dogs on the streets that became a public relations problem as the Sochi 2014 Olympics approached. Russian officials hardly wanted Olympic tourists and visitors to see tens of thousands of animals on the streets. They thus resorted to mass euthanasia campaigns; Moscow alone spent 1.3 billion rubles ($45 million) on dog shelters, sterilization and other programs between 2008 and 2009, but to what advantage remains unclear.[42] As for Sochi, dogs disappeared en masse, and most were likely euthanized, perhaps up to 2 million of them.[43]

As for the broilers, the Russian Federation's legislation is in concert with that of the EU, although government agencies and animal activists have been much less energetic in verifying standards. According to the FAO, Russian statutes contain "a number of provisions designed to safeguard the welfare of poultry, including requirements concerning stocking density, feed and water, litter and ventilation." Russian laws are, however, less comprehensive than those of the EU, with no species-specific legislation on the welfare of pigs and beef cattle, and "a number of shortcomings in the welfare of poultry at slaughter."[44]

Existing legislation requires that owners of animals must provide them with feedstuffs and water that are safe for animal health and the environment and that comply with veterinary and sanitary requirements, with the use of certified feeds and additives, and premises vaguely "favorable" for animal health. For chickens, the stocking density for outdoors yards per 1 m^2 is 11–12 young birds and 3–4 adult birds. The statutes paid attention to water and feed, flooring, ventilation, cleanliness, cage size, safety, ventilation, temperature and light control according to international standards. Unfortunately, as recently as 2014, legislation introduced to ensure safety during transport, the stunning and slaughter of animals and so on does not appear to have been passed, and, while Russia is a member of the Council of Europe, it has not signed the European Convention for the Protection of Animals Kept for Farming Purposes or those conventions for the protection of animals during transport and slaughter.[45]

According to one study, Russian farming practices are comparable to Western practices, "in particular because there are many farms currently being developed and modernised with the help of European or American companies." A large number of small producers control the market in the Volga and Siberia regions, whereas export-oriented production takes place in the central and northern region. There is little vertical integration; slaughter plants and processors buy their animals from local farmers or they import animals. Veterinary monitoring seems present, not extensive, and there are few means to improve product quality. Management at many facilities seems inadequate because of a low "level of herd-management skills" even in the presence of specialized staff on the farms. Feed is often of poor quality, and the level of concentrates is too high. The study noted that "poor quality roughage and poor disease prevention programmes contribute to these problems." Judging from my brief, coincidental observations, transport takes place in typically ancient and rudimentary Kamaz trucks of Soviet design. Finally, the study noted that "laying hens are kept in battery cages, and broilers live in large barns. Both are comparable to the systems used in Europe."[46]

This inconclusive evaluation of Russian standards has been confirmed by *Poultry World News*, which noted that just four countries – New Zealand, the UK, Switzerland and Austria – received the highest

"A" rating in the Animal Protection Index of the UK-based World Animal Protection (WAP) organization. WAP noted that "some of the EU's largest meat-producing countries, including Germany, Denmark and the Netherlands, where animal welfare issues are given high priority by government," had very high standards, with those in Chile, South America's top-ranked nation, on a par with these EU member states. But North America – mostly because of the US industrial practices – received only a "D" grade, and China, Russia, Egypt and Vietnam bleated at "F," while with a "G" were Belarus, Iran and Azerbaijan where specific animal welfare legislation was virtually absent.[47]

Protest from the Restaurant Industry

Restaurant chains that purchase vast quantities of processed cuts, ground flesh and mechanically treated protein fiber have begun to sense consumer concern over health and safety, not only price. In 2015, Chipotle, a chain popular among young clientele, removed pork from its menus over concerns about "animal welfare issues," although its spokespeople did not single out which processors were guilty. The Chipotle website indicates owners "committed" to food safety in a way that may pressure big meat operations.[48] Target, McDonald's and Campbell Soup have all announced they will eliminate gestation crates from their pork supply chains.[49]

Most meat served in chain restaurants is produced in factory farms that, depending upon the country, use antibiotics to a greater or lesser degree to accelerate growth and prevent disease in animals. Yet, as noted, overuse of antibiotics promotes dangerous resistance to such infections as MRSA (staph.), a major threat that the US Centers for Disease Control (CDC) have signaled has become a leading public health concern. In a report sponsored by a half-dozen active international NGOs, specialists documented the challenges in removing antibiotics from the food chain. Hence, in addition to lobbying governments to introduce stricter regulations, they have also conducted public campaigns to encourage restaurant chains to abandon factory-sourced meat for more humanely – and safely – produced meat. They argue this is necessary given that Americans spend nearly one-half of the money devoted to food on meals outside the home.[50] Several chains (Panera

Bread, Chipotle Mexican Grill, Chik-fil-A, Dunkin' Donuts and McDonald's) have adopted policies to limit or prohibit these chemicals in any meat that they serve, while only Panera and Chipotle "publicly affirm" that the majority of the meat and poultry they sell "is produced without routine use of antibiotics." The others have established timelines to meet these targets, while Subway, Wendy's, Burger King and Kentucky Fried Chicken (KFC) have failed to do so.[51] The NGOs hope an ongoing campaign to identify misuse will end or significantly lessen the practice. They urge the restaurant companies to help in encouraging factory-farm companies to improve practices. They note that the economic costs of treating at least 2 million antibiotic-resistant infections, with some 23,000 deaths annually, are on the order of $55 billion.[52] For its part, the FDA has provided only "voluntary guidance."

A Slowly Growing Rumble: The Social Costs

Shoveling shit is a shitty job, especially because the CAFOs make you, the contractor, responsible for it. But it is hard to argue with a CAFO when the contracts you sign include a Kafkaesque non-disclosure that will hardly let you talk about their stipulations and requirements openly. But that has happened in the CAFO business. According to the Pew organization, some contracts have prohibited the growers from even acknowledging the fact of a contract with the company, and "some growers report not receiving copies of contract documents prior to signing or taking out loans." Only in 2009 did the USDA issue rules that required that these 21st-century indentured servants have written contracts, with the right to discuss contract terms with financial and legal advisers, business associates, direct family members and representatives of a federal or state agency. The contracts generally give integrators wide range of power – in particular, the power to pay poverty wages and to cancel a contract with wide discretion.[53]

Indeed, as with most contract labor in the US, contractors get few benefits – they have low wages and usually no pension, health insurance or workmen's compensation – and few in this industry can make a living solely from the broiler business. Thus, as noted, a number of them sell shit, er, fertilizer. According to Pew, in 2001, 71 percent of growers "whose sole source of income was chicken farming were living below the poverty line." Obviously, they cannot pay for proper waste management, but the CAFOs are not responsible legally for this. The workers are at the mercy of each company's "changing requirements for barn size, ventilation, watering systems, and other equipment," and the companies obligate "growers to pay for these costly fixed assets." Heavy debt and poor attention to the environment may result.[54] The workers are no different from chickens in their exploitation and poor conditions of life.

Big agriculture is like any other big industry from the point of view of race and ethnicity. Migrants, minority people and indigenes are disproportionately tied to these seasonal and difficult jobs; the rest of society benefits from cheap food. The migrants not only receive low wages and poor housing, but often have no social services such as health insurance, and even their children are denied access to public (in the US, state-supported) schools.

Poultry production takes advantage of the fact that, similar to agriculture and meatpacking, it is a low-wage industry that involves hard, exhausting work, high injury rates and miserable work conditions. Thus, processing facilities and factory farms have a difficult time in securing and retaining a stable force; there is high labor turnover.[55] For these and other reasons, the poultry industry often depends on migrant workers – some of whom, in the United States, lack documentation. Often, poultry farms are in counties with low population densities, low per capita incomes, and low levels of education, and this makes it easier for the migrant workers to avoid the scrutiny of immigration officers, and for factories to hide safety issues from the Office of Safety and Health Administration. Plus, the farms are normally located in "non-union" states, so that there is no one to protect the workers.[56] Gainesville, Georgia, known to us as one of the birthplaces of CAFOs, has a large number of migrant workers, a majority of whom lack legal immigration documents – and many from Mexico, a large number of whom work in the poultry industry, which employs nearly 8,000 people, must live such a difficult and fearful life.[57]

Local communities and local groups have begun to reject industrial farming, even when the perpetrator is a member of that community. When one farmer journeyed to the dark side of industrial chicken farming in Gressey, France, a town of 550 inhabitants near Houdan, with the intention of building a hen operation to produce 35–38,000 eggs daily, the residents united in protest. They rejected the disruption of "the tranquility of the village," with some 40,000 *Gallinaceae* expected to arrive, living some 200 meters from the nearest homes. The residents attacked with a protest banner: "The Winds Will Fill Our Homes with the Smell of Manure Droppings!" The farmer responded that he had a right to work, too, and pointed out that local laws permitted his new installation: "'I'm not here to put people back,' says Jean-Luc Lecoq. 'I do my best to limit the nuisances but it is also my livelihood, my job.'" And how will he treat the droppings? They will fall onto a treadmill that leads into a dryer, so that only 10 to 20 percent of droppings will be produced outdoors.[58]

In some places, chickens still hold a special place in agricultural communities into the twenty-first century. This is because, unlike factory farms, village poultry do not require much investment. If commercial poultry production, like "green revolution" technology, needs large capital inputs of special animals or seeds, fertilizers and biocides and so on, as well as considerable technical skill and sophisticated markets, and if profit margins of commercial poultry production will become slimmer due to increased costs of better biosecurity, balanced feeds, infrastructure and transport costs, then village poultry are relatively inexpensive to raise and a good source of nutrition. In a number of African nations, family poultry production, which accounts for 80 percent of the poultry sector, significantly improves family nutrition and incomes, offers employment opportunities that often promote equity for women, and is a buffer in times of crisis. World Bank projects reveal that small poultry farms can be competitive economically, and socially important.[59]

Chickens in villages also often help provide for people at risk, suffering from poverty and hunger – especially women, who are marginalized – or in times of war and civil war, or because HIV has decimated the labor force. Researchers have noted that family poultry development is ideal for rehabilitation of refugees and victims of disasters and wars. Somali nomads, who lost most of their cattle to drought, accepted poultry and poultry products as substitutes for cattle and beef. In households where there is a lack of able-bodied workers, such as those affected by HIV/AIDS or those with a disabled family member, village poultry offer nutrition and income without requiring much in the way of labor or financial inputs. Given that women are often the main caregivers for sick people and that chickens are usually under women's control, then chickens can provide them with additional resources.[60]

Indigenes may see their entire food system interrupted by industrial development, paradoxically in the name of animal rights – to protect species that are endangered, or megafauna mammals like whales that folks in cities do not enjoy seeing harvested. In these cases, the indigenes are forced into the distant market / supermarket / industrial food production chain – and into the consumption of factory-farmed animals. To adopt a culturally relativist argument, it would thus seem that cruelty to animals is recognizable in a variety of religious and other traditional practices – for example, ritual sacrifice – where those who carry out cruel acts can reasonably claim that the acts are protected by religious beliefs. And yet colonization of one people by another leads to the effort by colonizers to end what they perceive as abhorrent practices ("sacrifice" or hunting of endangered species), while they themselves slaughter millions of animals in a carnival of industrial bloodletting in factory farms.

To take a concrete example, Native Americans and First Nations people who subsist on fish and wildlife have struggled against the different rules and practices of colonists regarding how, when and in what ways indigenes might be hunters and gatherers.[61] Alaskan natives have the same problems, limits and confusions, even as they consume wild animals and fish for subsistence and from tradition. Since the 1960s, they have faced the construction of a pipeline across their lands and forced resettlement without legal protections. The Alaska Native Claims Settlement Act (ANCSA) of 1971 was intended to protect Alaskan Native claims and provide compensation, yet it ended all aboriginal hunting and fishing rights against a patchwork of worries, laws, regulations, and tensions between urban and rural residents, leaving Alaskan natives at risk of losing access to animals, fish and their habitats. Even more, industrial encroachment on their lands that has destroyed wildlife and dams that ruined fisheries create another challenge to preserving past practices, especially since newly imposed notions of sustainability and environmentalism have created obstacles to Native Americans and First Nations people continuing hunting.

The Government of Manitoba urges that "people have a responsibility to ensure that their actions in taking fish and wildlife do not damage the local population of the resource taken. When the level of use is greater than the ability of the resource to sustain that use, the area can no longer meet the needs of the people. At times, restrictions become necessary for all fishers and hunters to ensure the survival of the wild populations."[62] Thus, to require people to drop their religious practices as a violation of universal animal welfare standards, while settlement and the encroachment of industry have already done violence to those practices, seems inconsistent. And, as the situation of broiler production indicates, it may be that carrying out industrial slaughter of billions of animals annually throughout the world in factory farms is no more defensible as a reasonable practice.

Those directly affected by CAFOs in the communities surrounding them, and the contractors, face significant quality-of-life issues. While my focus is on the chickens, it is still of value to consider the plight of people in those communities. Simply put, those who run CAFOs rarely, if ever, consider ethics as having a place in agriculture, have little interest in mediating conflicts between industry, farming and the public – including those over pollution and public health – and see the entire point of vertical integration and contract production as keeping costs down, not improving community life. Stanislaus Dundon has worked to see family-managed farming survive, while embracing a series of ethical commitments needed for it be a vigorous part of the United States food production system. He notes that people understand without question that medicine, law and other fields require ethical foundations. He asks: then why not agriculture? Cannot the production of food, fiber and forest products be conducted ethically? That was one of the ideals of the American republic, as Thomas Jefferson believed. Ignoring the plight of his slaves, Jefferson saw the self-sufficient agriculture of hard-working, virtuous citizens as morally superior to burgeoning urban industrialism. An agricultural nation might avoid the pitfalls of vice that plagued industrial life.[63] Jefferson could have had no idea how agriculture in the twentieth century would become such an industrial activity – and one of such "vices" as disrupted communities, pollution and threats to public health. But, in factory farming, the agriculturalists of the world are being informed by food merchants and industrialists, big pharma and policy makers that agriculture is science, technology and production alone.[64]

The Pew charities have engaged the dilemmas of the growing labor, community and environmental crises of the CAFO. In one report, the former Governor of Kansas – himself from a family dairy farm in Saline county, Kansas – John Carlin, noted the transformation of such farms into industrial enterprises that entangled a variety of actors, involving small communities, private enterprises, corporations, federal and state regulators, and the public. Carlin claimed, "We care about the well-being of rural communities, the integrity of our environment, the public's health, and the health and welfare of animals." But he worried about the evidence that the industrial farm brought about tremendous increases "in short-term farm efficiency and affordable food, but [that] its rapid development has also resulted in serious unintended consequences and questions about its long-term sustainability."[65] Unfortunately, the work of the Pew Commission on Industrial Farm Animal Production was slowed by the fact that precisely the factory-farm industry was not fully forthcoming in analyzing the social and environmental costs of CAFOs – indeed, it slowed the work, with some representatives responding with "open hostility," in fact discouraging some authors from cooperating with the Trust, and even threatening to withhold research funding for their college or university.[66]

The Pew Memorial Trust raised significant concerns about the social, environmental, political and economic impact of factory farms dozens of years ago, regarding the displacement and disappearance of the diversified family farm based in communities with local people, local suppliers and local bankers, due to factory farms focused on one highly technical, specialized animal-meat operation. These factories, operated by large corporations of absentee landlords, had more concern about profits than about the integrity of rural communities, the impact on public health in the towns surrounding the farms, and the failure to provide health insurance, workmen's compensation or pension support to the laborers. In factory farms, not surprisingly, we care little about fowl, and almost as little about those who raise them.

6

Drumsticks

> I was eating in a Chinese restaurant downtown. There was a dish called Mother and Child Reunion. It's chicken and eggs. And I said, I gotta use that one.
>
> – Paul Simon

Chicken took off in flight on their measly, engineered wings to international markets in the 1980s. By the 1990s, chicken was "transnational," an overused and misconstrued term, often offered interchangeably with "international," but one that indicates activities of corporations and social groups that take place in more than one country. Largely supported by their countries' governments through direct and indirect subsidies, scientific R and D, and friendly regulation, and augmented by multinational corporations, the chicken trade grew from about 10 billion chickens in 1980 to 20 billion in 2010 and 22.7 billion annually in 2016. The trade involves whole birds (skinned, quartered, deboned and otherwise presented), parts (wings – that come in at least four different products), breasts (again full, split, skinned, boned and so on), feet and "paws," and legs (with thighs, drumsticks and so on) – pieces of birds that require histories in themselves. Let us also mention MSM (mechanically separated meat, the bits and pieces of meat, with gristle and small amounts of bone that are used in chicken nuggets, sausages

13 *President Vladimir Putin of Russia with a chick at the agribusiness exhibition "Golden Autumn," December 18, 2014, in Moscow. Russia is a major player in the international trading of broiler meat.*

and hot dogs). Such countries as the US prefer white meat, but during big televised sporting matches consume literally billions of wings. Such countries as China enjoy feet. Restaurants are big consumers, especially of breasts. And apparently more and more people are fans of chicken and turkey meat frankfurters as opposed to those made of pork, pork and beef, or kosher beef, with or without nitrates, nitrites and increased risk of colorectal cancer.[1]

These widely different products and tastes, and the growing numbers of fowl, both result from and contribute to international trade. And that international bird trade, while in some ways shockingly efficient (how *do* chicken parts get from Brazilian CAFOs to Chinese consumers so quickly and without spoilage?), is also environmentally questionable from several points of view. There's one carbon footprint of production and one of transport. Another cost is related to extensive and growing pollution of land, water and air from offal, droppings, greenhouse gases, mercury and other heavy metals, antibiotics and so on. And then there are Avian flu and other diseases that have become international as the outbreaks follow trade paths, even with national and international biosecurity measures in place. This chapter looks at the most modern aspect of the modern chicken: its transnational life, if such a thing exists, in an international world, as a commodity, a measure of taste, and even as a tool of foreign policy. It offers discussion of how the EU, China, the US, Brazil and Russia have become major bird actors.

Russia is an interesting case in its hunting and pecking of the chicken as a product in high consumer demand over the last 100 years. Before the Revolution, chickens helped Russian peasant women make ends meet. After the Revolution, the country suffered through periods of hunger and famine, at least through the end of the Stalin period, largely because of state policies that sought to harness peasant agriculture to a great industrialization program – and led effectively to war on the Soviet countryside. Chicken was not an important commodity, but such others as grain suffered through the vagaries of drought, traditional agricultural methods and political pressures for the peasants to join in the revolutionary endeavors. Because of food uncertainty, on several occasions the United States helped Soviet Russia in times of crisis: in response to the 1891–2 famine; the 1918–23 famine and epidemics, especially in the Volga River basin after the Russian Revolution, through the American Relief Administration run by engineer and later president Herbert Hoover; through Lend-Lease during World War II to provide materiel – and again food – in the life-and-death battle against the Nazis; and in the 1990s, during the economic and food crisis after the break-up of the USSR. President George Bush sent Russia millions of chicken parts – legs and thighs – that were surplus in the US because of white meat preferences What to do with the legs? Bush and American farmers had the solution. They became for Russians "Ножки Буша" (*nozhki Busha* – Bush's legs). The sales of Bush's legs fed Russians who were intrigued and happy with the inexpensive food, and also permitted the US export industry to get a leg-hold in the Russian market; many Russians actually remember only the leg-hold, not the sating of hunger, and many of them criticize American agriculture for having taken advantage of Russia during a crisis.

The National Chicken Council estimated that Russians consumed 1 million tons of chicken from the United States in 2001, about 1.28 billion legs – that is, one-eighth of US production and one-fifth of US chicken exports.[2] In the early 2000s, as the Russian economy recovered, Russian farmers and President Putin began to claim that US chicken was unhealthful, consuming high levels of antibiotics and steroids, and they also rejected, along with the EU, US practices of washing meat in chlorine. It may have been that the chicken legs story was humiliating to the nation that had put the first man into space, but whose science, technology and everyday life had fallen on such hard times. In any event, Rospotrebnadzor, Russia's consumer-health agency, began a campaign against the imported US chicken meat as unsafe. Instead of Cold War, food wars commenced. The Russian government shut down several McDonald's for health violations in August 2014, a few months after the Russian annexation of Crimea, but actually in a tit-for-tat over sanctions that the EU and US put on Russia for its military actions against Ukraine. Whatever the disagreements between American chicken farmers and Vladimir Putin, free trade was just one issue, as overtones of worsening Russian–American relations were paramount in the food war.

Chicken production is both capital-intensive, with a variety of construction, energy, chemical and biological inputs, and highly profitable for owners, if not for contract workers. It is scientifically complex for its reliance on hybridization, genetics and careful study of light, feed and other influences over bird fattening – and a relatively simple endeavor involving feeding and housing animals. In those cases where it has become a major export business, state power, environmental regulation of a stronger or weaker degree, scientific research and massive investment came together to create chicken powerhouses. Yet these powerhouses have also run into major challenges in safety, waste and disease management as they pursue international markets. A series of scandals have rocked some of the national industries, and AI and other diseases have spread rapidly, under the not-too-watchful eyes of the industry, like foxes in henhouses, leading to loss of human life and the decimation of entire flocks of birds.

This chapter looks at trade in broiler chickens, and how domestic and international considerations influence that trade. The chicken powers – Brazil, the United States and the EU – have great influence here by virtue of their millions of tons in sales. But they must also deal with local and national concerns over tastes, safety, tradition, evolving markets and biosecurity issues.

Flying the Coop

Chickens can fly, but not very far, nor for very long: perhaps 10 to 15 meters and a few seconds. But they fly and ship thousands of kilometers around the globe in foreign trade. At any one time, approximately 20 billion chickens inhabit the earth, but humans slaughter 50 billion of them annually. In 2002, the United Nations FAO estimated that there were 19 billion chickens in the world, with China having the largest number, followed by the US, Indonesia and Brazil. By the mid-1980s, the international protein or meat trade, including poultry, had begun to explode, leading to a series of fraught relationships among producers, competitors, trade partners and consumers over the nature of highly vertically integrated industries, international standards – for health and safety to AI, BSE and other diseases – and trade wars.

Birds roosted both at home and at restaurants. Retail groceries sold 58 percent of the product in 2005, with food services accounting for the rest, and with such fast-food outlets as KFC, McDonald's and other international food brands accounting for 60 percent of the product, or 25 percent overall. Indeed, for many McDonald's joints, chicken in strips and sandwiches and nuggets outsell burgers. Subway, a maker of sandwiches, is moving into the larger markets. When it comes to other restaurants, it seems that chicken sold in Caesar salad is the most common item, followed by strips or tenders (goujons), and chicken noodle soup.[3]

In 2010, global poultry-meat production was 97 million tons. In poultry meat production, the US, China, Brazil and the EU-27 ranked in the first four positions, with the EU share 12 percent, and the largest exporters, the US and Brazil, contributing 56 percent of the global exports. Other large producers were Russia, Mexico, India, Argentina, Iran and Japan. The ten leading poultry-meat exporting countries share 88 percent of the total export volume. The leading importers of poultry meat are China, Russia, Japan, Saudi Arabia and some EU countries.

If one considers the 21st-century chicken a technological commodity, then one can understand that Brazil and the United States became leaders in this modern technology, as they had in such other products as automobiles, airplanes and the like, in the usual fashion: a close relationship between business, government, and R and D, in either universities or government-based research centers; the presence of politically powerful agribusinesses that can command cheap labor and economies of scale in production facilities; large amounts of relatively inexpensive land; seemingly low costs for each chicken unit; through genomics, selection and hybridization, the creation of units that fatten quickly; and the willingness to ignore or downplay such externalities to production as pollution and the amorality of the mass-production killing machine that international trade in meat has become. In the case of chicken units, the belief that chicken meat is healthier than red meat, not least because of BSE, also gave chicken meat a push in international markets.

Before Brazil's remarkable rise to the spot of leading poultry producer and exporter in the world along with the US, the EC (before 1993, the EEC) had been the world's second-leading poultry producer and the leading poultry exporter. In 1990, EC poultry production totaled 6.2 million metric tons, or about 17 percent of the world total. EC poultry exports totaled 1.1 million metric tons in 1990, or about 42 percent of the world total including intra-EC trade, and 25 percent of the total excluding intra-EC trade. France, followed by Spain and the United Kingdom, was the leading producer. The EC poultry industry is comparable to that of the United States in structure, production technology and efficiency – generating 1 pound of meat from 2 pounds of feed – but has significantly higher production costs, particularly for feed and labor, the latter because of higher wages generally and social programs including full medical benefits, pensions and other important support. (In 1986, broiler production costs were about 35 cents per pound in France and 55 cents per pound in the United Kingdom, compared with about 30 cents per pound in the United States.)[4]

Total global beef, pork and poultry exports for 2002 rose to a record 16.7 million tons, up 7 percent from the previous year, with Russia, China and Japan the major markets that increased their imports. The opportunity here for the leading producers was clear because of their low production costs and high output, especially in poultry, with the US and Brazil competing for market share. The US market share around the turn of the century was about 29 percent of total meat exports. For 2002, total poultry trade reached a record 7.2 million tons.[5] How did Brazil come to rival the US?

Brazil

The Brazilian poultry industry was nonexistent until the post-war years, with chicken produced mostly by small farmers, and rarely did fowl make it to distant markets, facilitated by middlemen. In the 1960s, under a military junta determined to modernize all industry and society, agriculture benefited from investment, federally supported R and D, and urbanization that brought together demand, growth and science, eventually to create modern broilers raised in CAFOs. The development of the poultry industry, according to its publicists, also triggered a social revolution in the sector, in which the production chain by the twenty-first century provided direct and indirect employment for 4.5 million people, encouraged poultry farmers and their families to settle in new regions, belatedly fulfilled a dream of the nation's leaders to settle the country's interior, and provided economic security to small Brazilian cities. By 2007, Brazil exported 3.3 million tons of chicken to 150 countries.[6]

Chickens have been in Brazil since around 1500, and the first breeding farm appeared in the late 1890s in Minas Gerais state; gold deposits discovered at the time in Minas Gerais generated great demand for high-protein food for mine owners and miners. Also stimulating the chicken industry was a threefold increase in the Brazilian population, from 10 million to 30.6 million people, from 1876 to 1920 – chicken was important in diets. Unfortunately, the available breeds took six months to reach a 2.5 kilogram weight. A crucial breeding farm, East Basse-Cour, belonged to Delgado de Carvalho, the first President of the Brazilian Poultry Society (Sociedade Brasileira de Avicultura, founded in 1913 in São Paulo to modernize the industry); later, the farm became the Rio de Janeiro Poultry Station where Orpingtons imported from Britain were adapted to the Brazilian climate. The farm also studied Plymouth Rocks. Even after the station opened, breeding lagged, and faster-fattening birds were more widely available only in the 1930s.[7]

The arrival of European immigrants and their descendants contributed to the modernization of the Brazilian economy generally, and to the expansion of the poultry industry specifically. Many immigrants settled in states where poultry farms were concentrated – Minas Gerais and São Paulo, then Rio Grande do Sul, Paraná and Santa Catarina, and then Mato Grosso and Goiás. Here, Italians, Germans, Swiss, Ukrainians, Dutch, Swedes, Portuguese and Japanese brought their agricultural knowledge – and their fowl – to bear on the environment. The first breeder hens (Leghorns) were imported from Japan in 1937 and were used by the South Brazil Central Agricultural Cooperative.[8]

As elsewhere, World War II drove greater domestic production of foodstuffs, and in Brazil as well the Ministry of Agriculture decided to promote poultry. It built a poultry breeding station at the National Agronomy School in Seropédica, and it organized the National Poultry Cooperative to create other cooperatives. In the 1950s and 1960s, industrial production and processing techniques, including feed and health measures, entered the industry, followed somewhat later by genetics. The market included slaughtered and eviscerated birds. The poultry slaughterhouses doubled for other, smaller animals, and the facilities were simple, dilapidated, poorly arranged and primitive. But the government managed to attract foreign companies to invest in Brazil, and several birds that fattened more rapidly than domestic birds were imported from Pennsylvania. In the 1960s, Alfredo Rezende established a breeding farm in Uberlândia that became a one-day chick producer with 50,000 Arbor Acres breeders.[9]

Before the 1960s, Brazil was fully an agrarian nation; in the late 1940s, 70 percent of the population engaged in subsistence farming. The industrialization programs of the 1950s led to mass migration to cities. American breeder hens and foreign advice gave impetus to the industry, similar to the way in which the US hydroelectric industry contributed to the development of the Brazilian industry, especially with the engineering expertise of the Tennessee Valley Authority. Industrial poultry farming took off in the 1960s, at first at the pace of hunting and pecking in the absence of investment capital. But after the military junta seized power in 1964, the government created a National Bank for Economic Development (today's National Bank for Economic and Social Development – BNDES) that established an Agri-Industrial Reconversion Fund (Fundo Agroindustrial de Reconversão – FUNAR), both of which were intended to jump-start the development of the nation's massive interior – and secure a capitalist foothold against feared (and non-existent) Marxist insurgencies. This funding strengthened the poultry industry, as did the formation of the Brazilian Poultry Union (UBA), which worked to modernize production with new facilities and technical research in genetics that is a foundation for today's many successes.[10]

By the 2000s, over 90 percent of Brazil's poultry production was vertically integrated, and production rose from 500 million units in 1989 to 6 billion in 2008.[11] As in the US, the Brazilian broiler industry is geographically concentrated. Roughly 80 percent of chicken production is located in the South and Southeast regions. Santa Catarina state is the country's largest chicken producer and exporter. Chicken production – unlike beef, however – is far from the Amazon biome, and thus its environmental impacts are not felt there so much, in terms of rainforest deforestation, although there have been significant problems in chicken states in terms of manure, runoff and deforestation.[12] Working together, agribusiness companies, industry associations and the government have pushed capacity building and CAFOs; they offer a kind of extension service of talks and seminars on pre-slaughter and environmental management, poultry transportation, biosecurity and animal welfare. And, as everywhere else in the world, they assume that higher densities of birds in production can coexist with bird welfare.[13]

By 2015, Brazil's poultry industry exports totaled over $7.6 billion, or 4 percent of total Brazilian exports and nearly 9 percent of Brazilian agribusiness exports.[14] In 2016, the Brazilian production reached nearly 13 million tons, about two-thirds of which went to domestic markets; per capita consumption was 41.10 kilograms a year (or 90 lb, or 4 ounces per day). The chicken industry grew on the foundation of a public–private partnership that combined state-directed R and D in agriculture, subsidies to the private sector, and pursuit of growing world demand for poultry meat protein. The industry is highly technocratic, using R and D that improved genetics, and employing highly efficient handling and production facilities. The industry produces: whole chicken breast, skin-on and skinless; half-breasts, boneless and skinless; legs, with and without skin; MSP for use in nuggets and hot dogs; feet; skin; gizzards; hearts; even paws, knee cartilage, neck; half-chickens; wings; wing tips; "tulips" (with the meat flipped "inside-out" to expose the bone and then cooked and/or seasoned); drumsticks; whole grillers, boneless or in nine pieces; and the sausages and franks themselves, too. This industry is not afraid to play chicken.[15]

As in other countries, trade and other business associations represent fowl producers in the efforts to secure reasonable regulations and subsidies, and to facilitate expansion and trade. The Brazilian Animal Protein Association (ABPA), the largest organization representing the animal protein industry in Brazil, with over 135 members, came into existence from the union of the Brazilian Poultry Association and the Brazilian Meat Producers and Exporters Association. The ABPA claims sustainable agricultural practices and a close union with environmental organizations. Indeed, according to the Department for Environment, Food and Rural Affairs of the United Kingdom, each kilo of chicken in Brazil is produced with half the CO_2 emissions of British production: 1 ton of chicken manufactured in Brazil emits 1.2 CO_2-eq/ton, while in the UK it is 2.2. Granted, because of distances to transport the product within the country and for export, the Brazilian CO_2 emissions rise to 2.57 CO_2-eq/ton, but still less than the UK 2.82. ABPA claims that "Brazilian agribusinesses are located away from the Amazon biome and count on preservation projects for environmental resources. These industries also maintain water conservation projects and special treatment in closed loops. In this respect, Brazil has helped the world by 'exporting' water to locations that need it."[16]

Yet not all is well in the Brazilian export kitchen. On the one hand, Brazilian authorities insist that the industry meets international standards for health and welfare. ABPA, the Ministry of Agriculture, private companies, state-level poultry associations, universities, EMBRAPA (the Brazilian Agricultural Research Corporation) and the World Society for the Protection of Animals (WSPA) drafted a 50-page report (2008) to ensure health, safety and biosecurity from housing and raising, to delivery and slaughter; maintenance of facilities; proper density; ventilation; temperature, lighting and water; and biosecurity.[17] Brazil has pushed its animal welfare efforts forward – in part, to remain a good trading partner; many other countries in South America, Africa and Asia have no national animal welfare legislation and no national control over animal experiments, whether within or outside research establishments.

The success in building the industry and preaching animal welfare policies must be weighed against scandals of bribery of officials to hide CAFO public health and disease failures. In 2006, when cases of Avian Influenza broke out in Asia, Europe and North America, a massive coordinated government effort monitored and controlled the situation.[18] Yet recent exposés show that something is rotten in Rio. The top Brazilian meat producers bribed food safety inspectors to obtain certificates for rotten meat. If a sharp decline in global commodity prices and low investor confidence had already hurt the industry, then the so-called Lava Jato (Car Wash) corruption scandal that extends from oil and gas and the top of the Brazilian government to food industries, also tainted beef and poultry with a foul stench. In addition to bribes, such major companies as JBS and BRF, the world's largest beef and poultry producers, respectively, sought to pass off spoiled products as edible by flecking them with acid in order to mask the "rotting stench." Major buyers the EU, Korea and China suspended Brazilian beef imports in the wake of the scandal, exposed through an undercover sting called "Operation Weak Flesh."

The Chinese suspension hurt the industry egregiously since China has been the top destination for Brazilian beef, accounting for about one-third of all exports at roughly $12 billion, and this occurred two years after the countries reestablished trade in 2015 following a previous food safety scare. Brazil's agriculture minister Blairo Maggi, the owner of the Amaggi Group, the world's largest soybean producer – a company that harvests, processes and exports soybeans, and owns soy terminals, highways and waterways – and an engine, through his company group, of deforestation who even defends deforestation as positive,[19] defended the quality of the meat, even if beef is closely connected with deforestation and comparatively low job creation, by trying to affix the blame on only a few rotten eggs. Maggi claimed that the responsible officials had identified and recalled containers carrying affected meat. To make the matter go away, Maggi announced that the government had singled out and shut down illegal cattle operations, an action that made the state a champion of sustainability.[20]

If production costs for broilers are slightly higher than in the US, then the Brazilian industry resembles that in the US in its organization. The Brazilian poultry sector has become vertically integrated and highly concentrated, with the top five companies accounting for three-quarters of production in 1986. In addition, the federal support for R and D has impelled the Brazilian chicken far beyond national borders. EMBRAPA, the federal farming research agency, one of the world's leading federal agricultural research institutions, has been an engine of state–private-sector poultry growth. EMBRAPA's Swine and Poultry, part of the National Research Center of the Ministry of Agriculture, Livestock and Food Supply, supports research, development and innovation for the sustainability of swine and poultry production. Founded in 1975 in Concórdia, Santa Catarina, the nation's biggest chicken exporter, the lab coordinates poultry research in Brazil in such centers as the Campinas Agronomy Institute (IAC), the Minas Gerais Company for Poultry Agribusiness (EPAMIG) and the Advanced Center for Technological Research in Poultry Agribusiness (CEPTA); with private companies; and with such universities as São Paulo State University in Campinas (UNICAMP), the Federal University of Vicosa in Minas Gerais, the Rio Grande do Sul Federal University (UFRGS) and the University of São Paulo (USP).[21] It focuses on control of diseases, improvement of animal feed, upgrading of the genetic quality of animals, and environmental issues.

Already, by 1990, Brazil had become the world's fifth-largest poultry producer, with 2.4 million metric tons, accounting for about 7 percent of the world's total industrial manufacture. Brazil was third in the world in poultry exports with 11 percent of the total. In the 1980s and early 1990s, a shift in world poultry import market demand away from Brazil's traditional markets in the Middle East (Iraq, Egypt, Saudi Arabia) to Asian markets (mainly Hong Kong and Japan) benefited competing exporters such as the United States and the EC, although Brazil quickly regained market share in the 2000s.[22] But the relationship between China and Brazil is special. Over the past 15 years, China's explosive growth has pushed Brazil into a global agricultural powerhouse. Driven by growing Chinese demand for commodities, Brazil went from net food importer to the third-largest exporter of agricultural products in the world. In 2009, China became Brazil's largest trading partner, leaving the US as a distant second.[23]

China

China plays a fascinating role as a chicken player in the world's, and its own domestic, markets, as the world's largest egg producer and the world's largest importer of meat; its rapidly growing CAFO broiler industry, despite the fact that it still relies heavily on small-vendor, small-farm producers of live chickens to be slaughtered at market; its product preferences, including chicken feet; and its difficulties in introducing biosecurity measures to fight against Avian flu and other epidemics. The remarkably rapid modernization of the Chinese food industry may be the source of some of these paradoxes of plenty and problems. In the 1990s, the Chinese poultry industry was domestically oriented and relatively unsophisticated compared with Western industries. Most Chinese poultry production was in household operations that marketed live birds in urban areas – hardly CAFOs. But, as in Russia, the growth of a relatively affluent middle class meant the transformation of production into large-scale industrial operations, facilitated in part by relatively low production costs (especially for feed and labor), and imports of quality breeding stock to meet growing demand.[24] China has become the world's fourth-largest poultry producer and consumer, accounting for 8 percent of total world production in 1990 at 3.2 million metric tons, about 80 percent of which was chicken meat, and the rest duck meat.

In the most difficult political stages of the Chinese revolution, including the economic recovery from the failed Great Leap Forward and the turmoil of the Cultural Revolution, the broiler industry grew quite slowly, only doubling in production from 1961 to 1978, as backyard poultry farming for subsistence prevailed. From 1979 to 1996, a kind of market system took over, with faster growth as commune production shifted to household production; production increased 4.5 times. The sector became more intensified, consolidated and integrated from 1997 to 2009 with the rise of standardization in feeds, breeds, medicine, production facilities, slaughtering and processing facilities, and food safety inspection, and since that time the American model of standardization and scaling-up has fully taken hold.[25]

Imports remain vital to an increasingly ravenous Chinese chicken public, even as domestic producers begin to meet demand with a modern broiler industry, especially in China's rural markets, and also such international markets as the EU and Africa.[26] The poultry industry involves the hatching and raising of poultry (chicken, ducks, geese, ostriches and quails) to produce both meat (the broiler industry) and eggs (the layer industry). China has about 1.5 billion layers in stock, with 1.2 billion layers in production. China's egg production accounts for over 40 percent of global production, with eggs exported to Japan, South-East Asia and Russia, and with exports growing almost 10 percent annually in the 2000s.

In the twenty-first century, in concert with urbanization and a growing middle class, poultry, meat and eggs have become major products, whereas before they were considered luxury goods for consumption on special occasions. Consumption of chicken has risen ninefold in the last few decades, to over 9 kg per capita annually. The growth of fast food, supermarkets and other icons of the 21st-century food world have also contributed to this huge increase. To meet the demand, poultry husbandry has rapidly industrialized, with intensification of farming and vertical integration, with a large number of firms competing over low prices with low profit margins, and thus immense pressure by firms to cut costs. Firms thus have turned to raising more and more birds in CAFOs.[27]

Rapid growing pains have accompanied this transformation, including issues of "biosecurity": disease, cleanliness and wholesomeness.

Avian flu outbreaks such as those in 2013 and 2017 triggered sharp drops in poultry demand (and perhaps led to scores of human deaths). Cost-cutting practices have generated serious worker safety issues as demonstrated by the fire that killed 120 people in the Jilin Baoyuanfeng Poultry Plant in 2013. These matters have led the government to become more actively involved in health and safety measures throughout the supply chain. In addition, consumers are changing the face of the industry. As shoppers turn more to processed chicken in supermarkets, the small traders and wet markets (selling live and dead animals) that may be contributors to the food safety problems will be displaced. Growing incomes in rural areas mean less need for small farmers to keep poultry to generate cash on the side. In addition, as in other sectors of the economy, technological change – in feed, chickens and organization – is contributing to growth, as are modernizing extension services, disease prevention and control, quality standards and information. The live market, ironically, is far from dead. It sells mainly indigenous birds that make up 20 percent of the poultry market, and this market continues to grow.[28]

Diet and demand mean that the Chinese poultry industry, both wet market and factory farms, is thriving, and intensification has speeded up. This has permitted such major companies as Tyson, Cargill and Brazil's Marfrig and BRF to integrate in the Chinese market. Such Chinese companies as Wens have adopted the integration model, and, through contracts with farmers, are producing close to a billion birds annually.[29] Indeed, Chinese capital helped to expand the industry in part through the acquisition of foreign companies whose know-how and technologies offer crucial impetus. Tyson, for example, now under the control of a Chinese company, is building up to 20 modern sanitary chicken factories in China. The Shuanghui International Holdings bought Smithfield Foods of Virginia for the Chinese pork market.

But the Avian Influenza (AI) crisis is far from over, and, together with a series of food-quality disasters, this has severely diminished confidence in the industry at home and abroad. Between February 2005 and May 2007, 18 of the 31 provinces in Mainland China reported cases of AI, including cases of bird-to-human infection and human death. There were at least 25 human AI cases, with 16 deaths by 2007. Guangdong Wen Foodstuff Group Co. Ltd., in October 2017, announced a switch to new broiler breeds and expanded sales of chilled poultry meat because of another massive outbreak of AI the previous winter that killed almost 300 people – and cut profits 63 percent. The authorities shut down wet markets; this decimated trade in China's native yellow-feather chickens.[30] Chinese researchers have developed effective vaccines that are provided free to poultry producers nationwide, and generally food security has improved,[31] but domestic and international confidence remains low. It does not help that Chinese farmers have been giving Tamiflu to their birds, which may mask the disease in some of them and permit spread to other, untreated chickens.

Other scandals have buffeted the Chinese chicken industry. KFC, China's most successful fast-food outlet, has been accused of using broilers in its products that are fed toxic chemicals that even kill "the flies that buzz around" them. The Chinese press reported that a poultry supplier to KFC used a growth-accelerant feed called "instant chicken" that will speed up the growth cycle to 45 days – considered standard for broilers in the EU and North America, one must recall. Yet the additives, produced by the Suhai Group, include silver nitrate, chloride and a sterilization medicament. On top of this, having turned to CAFO methods, the Chinese industry may pack in more chickens per shed than its American or EU counterparts, with greater risk of infectious diseases spreading. And this, in turn, may have led to the use of more additives and antibiotics in the feed.[32]

China's production of chicken meat will grow over the next years, relying more and more on white-feather, yellow-feather and hybrid broiler production. Although China's numerous AI bans on foreign suppliers "has created a shortage of new genetic material for its poultry flocks, China has compensated by widely using forced molting, importing more grandparent stock from New Zealand, and developing a domestic great grandparent breeding industry."[33] To take one example, Guangdong Wen bosses hope that turning to genetic science will enable them to provide wholesome chicken products to the fast-growing supermarket and fast-food sectors, drawing on an in-house gene bank and breeding new varieties of meatier chickens. This is a significant break from "Wen's roots serving local markets" – with wet market birds.[34]

Chickens and their parts cannot simply fly around the globe. They

14 *Two workers in a Chinese wet (fowl) market.*

are subject to the vagaries of bilateral and internal trade agreements, and worries about disease. Those who know the limits of President Donald Trump's understanding of any complex phenomenon, such as trade or nuclear weapons control, will not be surprised to learn that his jingoistic attacks on China's "unfair trade balance" with the US have also had an impact on chicken exports. In his fury to engage in a trade war that will raise consumer prices but not address the trade imbalance, Trump has forced Chinese officials to put anti-dumping and anti-subsidy duties on US white-feathered broiler chickens. Whom did this help? It did not help US producers, Chinese consumers or American farmers and consumers. When China's Commerce Ministry removed the duties, Trump failed to understand his actions had not had the desired outcome: the Chinese simply turned to the Brazilian chicken industry for its powerful appetite for chicken feet. The US had managed to open the door for Brazilian feet in Chinese mouths.[35]

Similar trade disputes with South Africa had long buffeted US – South African trade, before an agreement ended 15 years of short-sightedness in a battle over fowl dumping. In that case, US chicken had been locked out of a market by South African authorities who claimed that American poultry farmers "were 'dumping' chicken meat in South Africa by selling at unfairly low prices." An agreement of June 2015 settled the dispute by establishing an import quota. But the quota and difficulties in negotiating an agreement indicate that free trade in fowl with China, South Africa, France and many other countries remains a free-range dream.[36]

The Impact of International CAFO Trade on Local Communities

International trade in broilers has put immense pressure on a variety of communities in nations whose own industries cannot compete. Nearly 70 percent of Africans who live south of the Sahara work primarily in agriculture. Yet African farmers are the world's poorest, and, over the past half-century or so, agricultural performance in Africa has deteriorated. According to the International Food Policy Research Institute (IFPRI), from the 1960s Africa's share of world agricultural exports fell from 8 percent to 2 percent, and the sub-Saharan region went from being a net food exporter to a net food importer. In this time, many Western policy makers assumed a solution to agriculture problems in less-developed and poor regions would be found in "green revolution" technologies that improved varieties of crops, fertilizers and pesticides to match them, and farming methods. But if green-revolution technology helped improve outputs in a number of countries across the world, including India and China – for example, through new strains of rice and grain – in other places, its capital-intensive nature and high costs led to failure – in particular, in Africa and the Caribbean. Some of the reasons had to do with lack of infrastructure, even of the simplest kinds: canning and bottling facilities, feeder roads, refrigeration and so on.[37]

It is also a challenge to rejuvenate agriculture where markets are under constant pressure from producers in the US, Brazil and Europe, whose production costs give them great leeway to open markets. The example of the penetration of international chicken-parts trade into Ghana suggests in this light that CAFO chicken is an opportunity and a problem. As Pascal Zachary observes, in Accra's central market, domestic live broilers cost twice what imported Brazilian chickens do. The latter are frozen, but defeathered, cleaned and ready to cook. In Ghana, the chicken has become a symbol of global trade, like the other new foods that Ghanaians consume from this trade: Thai rice; Italian canned tomatoes; chocolate from Indonesia that is cheaper than Ghana's largest cash crop, cocoa; and so on. The problem is that, like many other agricultural nations, Ghana relies on exports, but lacks food processing and other infrastructure to export most things in a form that meets consumer expectations abroad. Zachary notes that wealthy European and North American nations, and such rapidly developing countries as India and China, have invested in farm productivity to raise the incomes of farmers, improve output and drive consumer demand for homegrown foods, but African countries do not manage to do so, and thus pay more for domestic food and rely to a greater degree on imported products.[38]

European chickens also abound – or peck around – in Accra, upsetting the market and putting Ghanaian farmers at a disadvantage. In 2003, about a quarter of European poultry exports went to Africa, but by 2018 nearly half of these exports went to Africa, much of it poultry of Polish origin. From 2010 to 2016, Polish chicken exports to Africa increased fourfold. When South Africa banned chicken imports at the end of 2016, European exporters quickly found other markets in Congo, Ghana, Gabon and Liberia. In Ghana, EU fowl is three times cheaper than local retail chicken, hurting jobs and an entire sector. But the Europeans ignored the contradiction in attitudes toward Brazilian birds; they wanted to sell Euro-chickens, but criticized Brazilian producers for hurting African farmers. As one journalist remarked, "They sell their chicken and do the ostrich policy."[39] One way to deal with *le poulet Européen* would be for Ghana to adopt restrictive import policies. Prohibitive tariffs might stimulate domestic production, but would probably be ineffective overall and impose significant costs on the consumer. In light of this, one study concluded that maintaining trade links with other West African countries and focusing on feed efficiency, through a mix of domestic production and imports, would benefit the national industry, offer better understanding of the potential for globally competitive chicken-meat production, and likely benefit other industries dependent on competitive feed, notably aquaculture.[40]

As the French ought to say, "All's fair in love and trade in chickens." The French chicken industry complains about competition from Poland, Bulgaria and Romania, and about the arrival of Brazilian chicken parts in Paris. Yet French chicken farmers seem not to care about the impact of their exported chickens on African concerns. African producers have accused their European counterparts of allowing cheap chicken legs to flock to the continent at the expense of local chains. They note that European farmers still "loudly denounce the free trade agreements with Latin America that . . . pose the risk of seeing Europe flooded with Argentinean or Brazilian meat," and yet "these same farmers do not hesitate to sell their chickens in sub-Saharan Africa, without worrying in the least about the damage done to local sectors."[41]

These stories indicate that, no matter the efforts of activists, environmentalists, ethicists, consumers and regulators to shape a profoundly more human poultry industry, the industrial imperative and global trade competition have pecked away at EU directives to achieve those ends. Indeed, AVEC, the Association of Poultry Farming, Industry and Trade of the EU, based in Brussels, tries to balance the desire to have a chicken in every pot with food security. AVEC members have a "vision for a future with European poultry on every table. AVEC is committed to improving knowledge, innovation and transparency in the European poultry meat sector. We champion the shared values of the sector so that every consumer, food professional, official and politician can have confidence in our products." Its members seek food security, as the "first responsibility is to consumers." They are determined to rear "healthy, thriving birds. Healthy birds provide healthy meat which is safe and nutritious." They take "antimicrobial resistance (AMR) very seriously, and AVEC is committed to minimizing the use of antibiotics." Yet the technological impetus or pressure comes from the encouragement of members in "the use of modern technologies to support farmers and poultry meat processors. Such technologies can give them a competitive edge on the market and provide greater market transparency and reliability." It is in seeking this competitive edge and in meeting (and generating!) expanding "high demand for European poultry meat outside the EU" that the tension appears.[42] It appears in Europe in rural areas, and it appears in nations with struggling agriculture – for example, in West Africa.

Russia: The Baba Yaga of Poultry Farming

If tastes and politics matter in the global trade of birds, then vodka-drinking Russia, wet-market China and backyard-raising Ghana have become enmeshed by the globally produced, factory-farmed broiler. This brief discussion shows that even countries that produce chickens in huge numbers in their own CAFOs still import to meet special niches and to share the wealth of bird meat according to domestic tastes and domestic challenges, including those of disease and the production of wholesome poultry. In light of this, what has become of Bush's chicken legs?

Baba Yaga, the archetypal Slavic supernatural witch with iron teeth, flies around in a mortar, wields a pestle, lives deep in the forest in a hut (*izba*), and stands on chicken legs. In her grotesque emaciation and with her wailing sounds, she suggests the archetypal Soviet chicken. On the eve of its break-up, the Soviet Union was the world's third-largest poultry producer and market, and was the second-leading importer. Yet it is also true that agriculture was the sore spot of the nation's economy. I bought a few communist birds in the mid-1980s before Mikhail Gorbachev's rise to power. They were scrawny and ugly, with bruises and feather follicles on unsightly skin. It was nearly impossible to buy cut-up chickens. or chicken of any sort wrapped in cellophane, although you might get fowl wrapped in *Pravda* or *Izvestiia*. If you bought chicken at a restaurant or dining hall, then you could be certain the meat had more joints, bones and cartilage than normally present in the *Gallus domesticus*, and the meat was inordinately tough.

If, in the 2010s, Russia was rebuilding its meat sector through substantial investments, then the prospects are uncertain for the 2020s. Russian meat producers face a number of obstacles to competitiveness and high-quality production, many of which date to the Soviet era's undercapitalized food sector, a sector that the state ignored or exploited for 70 years. The state pursued a murderous campaign to collectivize agriculture in the 1930s under Stalin that resulted in the peasants slaughtering roughly half of all livestock, and suffering through a famine in Ukraine and Kazakhstan that killed millions of people. Investment to rebuild the nation after World War II focused on heavy industry, even in agricultural regions of Russia, Ukraine and Belarus, which left agriculture without such basic machinery as tractors, and with poorly developed infrastructure of roads, electrical power, processing plants and refrigeration facilities. Leonid Brezhnev's Food Program (1982) could never overcome weaknesses in infrastructure, low-quality seeds, overuse of fertilizers, and an aging and disinterested farm labor force. To this day, Russia faces poor infrastructure (low production efficiency, poor feed quality, miserable feeder roads, insufficient energy delivery including electricity and power outages); outdated and inefficient production facilities; and changing government policies, strategies and investment decisions. Other problems facing Russia's broiler producers include disease outbreaks and food safety hazards.[43]

Although Soviet meat production reached an all-time high during the period of "Developed Socialism" under Brezhnev, whose food program centered on the effort to show the consumer that butter was as important as nuclear weapons, long lines remained the staple food of Communism. Mild winters, more land planted with forage, promises of leaders – none of these things either produced more food or cut the need for imports. Since Brezhnev promised meat, the Soviets turned to Eastern Europe, France, Argentina and New Zealand for beef, pork, chickens and lamb. This did not help, as meat consumption remained 25 percent lower than the official norm. But it means that US imports of Bush's legs in the 1990s were nothing new.

Agriculture being the sore spot of the Soviet economy, the poultry industry grew much more slowly than the expanding domestic market; imports filled the excess demand. Soviet poultry production totaled 3.4 million metric tons in 1990, or 9 percent of the world total. From 1965, the Soviet State Poultry Industry Trust, Ptitseprom, coordinated the vertical integration of the industry and accounted for the majority of Soviet poultry production and marketing in the state farm sector. From the mid-1980s, the Soviet government encouraged a shift to private production of poultry, but in 1985, still 65 percent of Soviet poultry production remained in the state farm sector. The Gorbachev economic reforms encouraged expansion of the private sector, and the collapse of the USSR ensured the final shift away from state chicken farms.[44] By 1991, Russia had lost its major export markets, too, when the Berlin Wall collapsed and East European palates salivated to the west, while the former republics branched out on their own.

In the early 2000s, as Russia began its recovery from the economic crises of the 1990s, largely through the sales of oil and gas, Russian officialdom and bird purveyors sought to rebuild the domestic food industry. In this environment, Russia claimed its status as the largest foreign market for US poultry (mainly chicken leg quarters) as Russian suppliers looked to the United States to fill the need for inexpensive meat protein. In 2002, total poultry imports were about 1.3 million tons. Russian bans on red meat imports from the EU and some other supplying countries early in 2001 stimulated poultry consumption. (It seems clear that domestic producers will never meet growing demand because of insufficient investment to upgrade production facilities, and feeding, processing and breeding systems,[45] so Russia will import foods of all sorts in large quantities for decades to come.)

Chicken legs have been a strange undercurrent of Russian–American relations since the 1990s. By late in the decade, chicken products were the focus of negotiations, debates and disagreements between the two nations about chicken meat imports, about additives and steroids, about invasions and about legs. A number of Russian officials believed that an embargo on Bush's legs would stimulate production at home, even if it sent local prices up. The populist mayor of Moscow, Yuri Luzhkov, even forbade the city government to buy any American chicken meat, claiming that American birds were full of steroids, the reason behind the epidemic of American obesity, and a growing problem for Russians since "these health problems are transmitted at a genetic level" to humans.[46] Patriotic protesters appeared in Pushkin Square, near the site of the first Soviet McDonald's – the location for historically significant gatherings of protest against Soviet human rights violations – calling on citizens to destroy the US and screaming "Long Live Russia!" in the effort to push a broiler embargo. Yet, in 2001, the Russian chicken industry was hardly pecking; representatives of 20 companies wrote to the Prime Minister Mikhail Kas'ianov and other officials with a request to impose quotas for a short time to permit Russian companies to become self-sufficient again.[47] This brief embargo ended after sharp US protests. Throughout the early 2000s, American–Russian give and take, threats of embargoes and quotas, chills over AI and so on characterized the chicken meat trade. Russian officials claimed incongruously that the embargoes would hurt the US housewife only, not the Russian consumer.[48]

Big-power politics played a role here, with the two cocks fighting. Putin declared in August 2008 that 19 American companies would be forbidden from importing chicken meat, while another 29 enterprises that did not correspond to the standards of Rosselkhoznadzor (the Federal Service for Veterinary and Phytosanitary Surveillance) would be added to the list, and he assured his citizens that this had nothing to do with US displeasure over the Russian invasion of Abkhazia during the Georgian conflict.[49] These were not only cockfights, however. Russian officials noted the fact that, like the EU, so Russia would forbid the import of US chicken washed in chlorine, with a threat to cut imports significantly.[50] This disagreement was briefly papered over with resumption of trade some months later, with the US promising to provide Russian health specialists with proof that US meat met Russian standards,[51] only to end in 2012–13 with significant tariffs placed on American meat, in part to help Russian producers: the annual quota was reduced from 857,000 chickens to 330,000, as part of protectionist policies. Still, "cheap chicken is delivered to the Russian market" from Mexico and Brazil.[52]

Federal policies that favored some businesses over others, and the general inclination of Russian capitalism toward rapid centralization led to consolidation of the Russian chicken industry by the late 2000s. This consolidation resembles that in other countries of the world, with larger agroindustrial businesses swallowing others whole, without chewing. With the acquisition of Kurinoe Tsarstvo, Cherkizovo has become the largest operator in the Russian poultry market, with a total output of 120–140,000 tons annually, or about one-tenth of the domestic poultry market, which in 2006 amounted to 980,000 tons. Each of the three next-largest poultry producers in Russia – the Prioskolie and Kochetkov poultry farms in Belgorod, and Severnaia near St. Petersburg – did not exceed 100,000 tons in annual production.[53] Roughly at this time, Russian chicken meat production finally reached the level of US imports.[54]

In the 2000s, during its recovery from the economic and political crises of the 1990s, Russia became the leading world poultry importer, buying about one-quarter of the world market's broilers and turkey. If Russia's market opened under President Yeltsin, then by 2001 the country already imported 1.4 million metric tons, or a whopping 82 percent of total fowl consumption. Working with the government, local producers pushed to relieve this dependence through modernization of industry and setting a limit of just over 1 million metric tons, with the US accounting for 73.5 percent of that total, a total that would grow to 1.252 million tons in 2009. [55]

Migration and Sale Without Passports or Medical Certificates

Global production – and export – of broilers will likely continue to grow modestly over the next years. There seem to be ample feed supplies at relatively low prices, and a lull in HPAI outbreaks will also help. It helps that broiler meat is competitively priced and easy to prepare, no matter what the economic conditions. Yet increasing prices for feed and transport, and future outbreaks of HPAI, must be expected to hit the market again because of climate change, growing competition, and lack of international standards and regimes to ensure biosecurity. According to USDA analysts, even China, faced with HPAI-related bans, limited availability of genetics, a saturated market and weakness in prices and demand, will likely see production grow, even if production remains below historical highs.[56]

Total poultry meat production in the EU-27 in 2011 was around 12 million tons, of which total broiler meat production was 9.6 million tons, harvested from 7.5 billion broilers. Seven broiler meat-producing countries in the EU have a production of more than 0.6 million tons each. The UK is the largest producer, followed by Poland, Germany, France, Spain, Italy and the Netherlands. Within the EU, the Netherlands dominates broiler meat export, with a share of 29 percent of total EU exports, followed by France and Belgium. Germany and Poland follow the leaders, but have been increasing their exports in recent years, and Ukraine has also joined the barnyard. As for intra-EU trade, the four leading importing countries are the Netherlands, UK, France and Germany, which account for 62 percent of all broiler meat imports. Other imports come largely from Brazil and Thailand.[57] US imports are not allowed because of insufficient safety standards.

But, elsewhere, US birds do well. According to the USDA, global chicken production reached over 90 million tons in 2017, largely in the chicken powers the United States, Brazil, India and the European

Union. US and Brazilian expansion has been driven by higher exports, while expansion in the European Union and India is due to growth in domestic demand. US production reached a record 19 million tons in 2018, with exports approaching 3.2 million tons.[58] Global exports are forecast to grow, perhaps more rapidly than overall production, with Brazil and the United States continuing market dominance, but with the EU, Thailand and Ukraine increasing exports, too. Brazil's industry will benefit from strong demand in its Middle East market. Hong Kong, Japan and the Philippines will increase buying, and such developing markets as Angola, Cuba and Ghana will fuel trade.[59]

You are what you eat. Americans prefer white meat, and restaurants and homes meet this demand. How do poultry factory farms make America great again with regard to dark meat, once they have separated broilers into their component parts? The US faces great competition in the parts business. From 2004, Brazil has been the leading chicken exporter, and this had an impact on US exports. US sales also ran into problems with Avian Influenza and ongoing turmoil in the Russian market. US exporters have hopes for continued exports to China, although it is unclear what the impact will be, in the short or long term, of Trump's specious trade war with China. Another market for the US may be African countries that seem interested in leg quarters. But the US will likely always fail to establish markets in the EU owing to higher standards of cleanliness and safety, and lower tolerances for additives and washes.[60]

The international technobird – the billions of such fowl ready to move from one country to another – have imposed an increasingly difficult task upon regulators and consumers worldwide. In light of the filth inside factory farms and the great risk of foodborne and zoonotic diseases, how can one ensure wholesome food products? How and when can inspection take place in the export world, who should carry it out, and where might weak points in the inspection chain appear? Do standards in one market or nation hold up to scrutiny in another nation or among trading partners? Differences in sanitary practices, feed additive controls, animal welfare standards, and disease monitoring standards and biosecurity measures – and such growing problems as the threat of Avian flu pandemics and national scandals involving unadulterated products sold as safe – indicate that, both at home and abroad, factory-farm safety, wholesomeness and public health are difficult to achieve. There is no simple determination for establishing crucial international standards for these categories of food safety based on some universal scientific language, as this brief exploration of Bush's legs and Brazilian feet has shown.

15 *The life of chickens with mother and daughter.*

Epilogue: Broiler Chernobyl

> It is better to be the head of a chicken than the rear end of an ox.
> – Japanese proverb

> Why did the Punk rock musician cross the road? He was stapled to the chicken.
> – Children's joke

The chicken you eat is almost certainly not a chicken in the traditional sense. It is a genetically formed meat machine, likely one of three models distributed by a bird genetics company, then produced in massive sheds by a large corporation – often, a multinational corporation. There is a one in five chance that the chicken meat you buy originated in another country. The boneless, skinless thighs you just bought – or the nuggets, in any event – certainly came from a factory-farm shed thousands of kilometers away, where it lived for six or seven weeks, with thousands of other broilers; where it moved about, a few centimeters at a time, bumping into other broilers; and where it gained significant breast and thigh weight from conversion of specially constituted feeds, often made from subsidized grains, with additives to promote growth. Still immature, the broiler was gathered up by hand or shovels or a chicken harvester that looks like a Zamboni – a hockey ice cleaning machine –

pushed, shoved or expelled into a box with dozens of other birds, and taken to a processing facility. At the processing facility, these broilers were hung upside down by the legs, strung to a conveyor, stunned in a variety of ways, and their necks slit to bleed out. Evisceration, slicing, dicing, cleansing, packing and so on followed. The low-paid slicers and dicers have miserable, blood-soaked, exhausting and dangerous jobs; they are the 21st-century Jurgis Rudkuses of the world – the central character of Upton Sinclair's *The Jungle*. The broiler chicken and machine now overlap in make-up and treatment in factory farms.

We have discussed the history of the broiler, its rise on the cusp of nature and technology, and we have considered the significant impacts of the production of industrial chickens on society, public health, farming, the environment and new disease vectors; our declining moral standing from the point of view of the technological nature of the birds themselves; and the belated worries about their welfare in industrial conditions of production, and the welfare of those who raise them for us. No longer a farmyard friend, a happy and active animal, the chicken is now industrial output with all of the social and environmental costs that come with industrial production. The broiler industry is a serious, competitive, dangerous and bloody business wherever it has been embraced – in Brazil, China and Russia, in the European Union, in Ukraine and in the US – that is usually hidden from public view. Yet the chicken appears to us in bloodless, cellophane packages, without even a hint of the piles of shit, chicken carcasses, feathers and offal that are major byproducts of broilers, as are phosphorus, nitrogen and heavy metals that enter waterways, fields, forests and other ecosystems.

What came first, the chicken or the cellophane? We have considered the various archeological, historical, political and scientific understandings of the modern broiler, a creature at the intersection of nature and technology, regulator and regulated, and farm and factory. The broiler is a human project in its entirety, and came after the egg. But it also came from a former family friend, a farmyard creature, and before that the borderlands of forest and field. The broiler lives for us, but its life has been constrained by a productivist imperative. It pays us back and overwhelms us with its bones, feathers, offal and crap.

In the postmodernist story by the Russian author Viktor Pelevin (1962–), "Hermit and Sixfinger," two chickens in the V. I. Lenin

Broiler Factory, one of which indeed has six toes, avoid slaughter by learning to fly, having exercised to strengthen their wing muscles, and they also succeed, at least temporarily, in preventing the slaughter of their comrades. Otherwise, heaps of straw, loud, frightening noises and smells of vinegar overwhelm and frighten them. Danger is everywhere, and there is no place to hide.[1] As Hermit and Sixfinger are disillusioned by the broiler factory, so am I by factory farms, but at least they were able to fly the coop at the end of the story.

The agroindustrial complex presents dangers to us all, to peasants in China and Brazil, contract workers in the US, small farmers in France, and to consumers everywhere. It produces food cheaply and safely, yet it is cheap because it passes along many costs to the consumer that he or she feels later, if never fully comprehends – for example, fish kills, polluted waterways, climate change; and it is safe because we are inured to thinking that a few hundred thousand gastrointestinal illnesses and a few hundred or a thousand deaths annually from food processing around the globe are not significant numbers, given the billions of people and the billions of animals they consume. But we are the food guinea-pigs. The questions are how to sustain fertility of the soil through conservation, not driving it to the ground, and to ensure local food security; how to produce healthy, non-toxic food; how to ensure good salaries with social support in rural regions; and how to respect the goodness of animals and the environment. Basically, we are asking how to produce nutritious food, and how to treat animals with dignity before they die to serve our appetites.

Today's factory farms still apparently worry about, and fear, honesty and openness on the cruelty, safety and cleanliness of their operations. They speak in Orwellian language about animal welfare; their representatives have never read Pelevin or Sinclair. Large food producers use breeding strategies and economies of scale to produce food at low cost and high profit. At each stage of their operations, including in nomenclature, they treat animals as objects of manufacture to accelerate processing, reduce costs of animals and labor, and avoid consideration of the brutality of the egg-to-cellophane industrial cycle. Almost without exception, like any business, the owners, their lobbyists and lawyers seek to lessen consideration of livestock as animals to avoid animal cruelty laws and to treat the byproducts of production – offal, manure and so on – as manageable and minor inconveniences. This is truly a machine with inputs – feed, light, warmth – dedicated to ensure rapid growth and uniformity in birds, themselves perfect for machine processing. The growers get the short end of the stick: they are responsible for the labor, the sheds, and the water, fuel and lighting; they stand on the cliff-face of the battle to manage manure and waste – and for all this, they get low wages. And consumers get cheap food, but that food often involves quite filthy utensils.

Environmental circumstances often dictate the origin and management of diseases in humans and other animals. The rise of factory farms presents unique challenges in this regard. Industrialized hatching, raising and processing of animals have created new opportunities for diseases to arise and spread globally, and not only among flocks and herds, but from animals to humans. Thinking about avian diseases thus requires consideration of the ways in which factory farms might be modified in the name of meat creatures and humans alike. These considerations necessitate a revisiting of chicken ecology in the twenty-first century, a place and time where bird pandemics can occur, and also of the welfare and public health implications of avian diseases.

Recently, two writers argued that the symbol of the modern era ought not to be the automobile or smartphone, but the chicken nugget; I suggest that symbol ought to be the mushroom cloud of nuclear tests whose dangers still hover over us. But why did they propose the nugget? The authors make compelling arguments that reflect issues covered in this book. First, they note how capitalism has organized hierarchies of human life, of power and money, and also – crucially, in this case – of society and nature. The boundary between the two has become "violent and tightly policed . . . with deep roots in colonialism." Since the rise of capitalist institutions of conquest, nature has been a place to serve conquerors who take the wealth from it, and control the water, forests and mineral wealth, with the concomitant effort to subjugate local people (indigenes). Animals, of course, were part of this transformation of nature into a commodity, with those that accompanied the settlers serving their "ecological imperialism" in bringing with them a host of other flora and fauna.[2] Commodities also became global in meaning and value as part of an "industrial grain-oilseed-livestock complex." By the late seventeenth century, a mechanical view of nature had replaced a more organic or holistic one, as people came to view nature as a clock or machine – in the words of Carolyn Merchant, effecting some kind of "death of nature."[3]

Patel and Moore indicate how capitalism, and the profit motive, drove humans to seek to use nature's bounty as efficiently as possible – in the case of food animals, pushing them through breeding and hybridization tied closer and closer to industrial production for greater yields, and leading ultimately in the twentieth century to our broiler chicken. Other consequences of this new hierarchy between nature and society – that led to the chicken nugget – include the destruction of the intimate relationship between people and nature. This involved new labor relations of slaves and workers serving owners of land to exploit crops, lands, forests and mines. Eventually, the factory became the locus of production, including the factory farm. These structures used science and technology to expand production, to support growing urban regions and placate workers with cheap food. Patel and Moore end their discourse on the chicken nugget by warning us that, in spite of the spread of this system globally, and the production of more food and food products than ever before imaginable, poor food and ill health are often the result – hunger and malnutrition remain overpowering. The environmental consequences of these attitudes toward nature – and of the kinds of products that result – have been significant, overpowering and irreversible. The energy and capital inputs required have contributed to global warming, as a result of which less and less arable land will be available, and agricultural decline will result.[4] All of this begs the question, is cheap meat even possible? Should it be pursued? Must the broiler be abandoned?

Such groups and organizations as Compassion in World Farming and the Pew Campaign to Reform Industrial Animal Agriculture have recommended crucial changes to factory meat farms. They include: "caps on total animal density; shared financial and legal responsibility for proper waste management between farmers and corporate integrators; monitoring and regulation of waste transported off concentrated animal feeding operation (CAFO) sites; [and] a requirement that all medium and large CAFOs obtain Clean Water Act permits."[5] There must also be an end to cage cruelty, as an immoral and outdated technology. These would seem to be reasonable minimum steps – with added costs, to be sure, but they are needed to guarantee the safety of the public and the environment. Likely, these changes could be adopted universally, among all broiler producers. Many countries have adopted these guidelines, and more stringent ones, without disruption in production.

The novelist Samuel Butler wrote "A hen is only an egg's way of making another egg." In the CAFO world, the hen must produce 20 times more eggs than she used to, and toward the ends of egg production and restocking breeders. Henry IV is reputed to have insisted that "I want there to be no peasant in my kingdom so poor that he cannot have a chicken in his pot every Sunday." But this is likely false news as his reign was marked by rebellion after rebellion – and bloodshed, too. Yet Herbert Hoover did want every American to have a chicken in his pot. Henry Ford, for his part, asserted that "Business is never so healthy as when, like a chicken, it must do a certain amount of scratching for what it gets," by which he meant that businesses should be willing, at an early stage, to take risks. The factory farm seems to have placed all of the risky aspects of its existence on others, but it certainly embraced Ford's American system of mass production and control of all inputs through vertical integration. As for the CAFO sheds, Frank Lloyd Wright famously said: "Regard it as just as desirable to build a chicken house as to build a cathedral." This has nothing to do with CAFO sheds, but it certainly has to do with Wright's effort to bring simplicity, natural elements and repose into his designs, in the hope that such designs as his "prairie style" houses would achieve harmony in nature. But factory-farm sheds are unnatural, even if a place of final repose for broilers. And, after all, "The sky is falling," the chicken said. But, in this case, the sky *is* falling: the factory farm, as constituted, has brought us to the abyss of dangerously efficient meat production with unappreciated human, community and environmental costs. We cannot walk on eggshells about this conclusion, nor be chicken about reaching it, nor chicken out about the need to do something. If this takes ruffling feathers, then feathers we must ruffle.[6]

Let us remember that a broiler CAFO is not a nest egg for saving a bit of money each week, and the chickens in the sheds hardly scratch out a living. With factory farming, we are putting all of our chickens in one basket – less than 1 percent of meat chickens now raised are free range. The inputs into a factory are not merely chicken feed. Fifty billion CAFO birds are not worth two in the bush. Chickens in factory farms do not strut their stuff, nor are birdbrains, nor are they permitted to grow into maturity. In the pecking order of human existence, factory farms must have a strictly regulated place. We rule the roost here, not agroindustrial approaches to providing food. They are cocksure about their industry, but some of it is a cock-and-bull story. Let's not play chicken.

Notes

Introduction: Egg First

1 Hatte wohl Hunger,
ass noch ein Hühnchen
mit meinen Händen
und merkte beim Hühnchenessen,
dass ich ein kaltes und totes
Hühnchen ass.

2 Compassion in World Farming, "The Life of: Broiler Chickens," CWF, January 5, 2013, at www.ciwf.org.uk/media/5235306/The-life-of-Broiler-chickens.pdf.

3 David Kritchevsky, "History of Recommendations to the Public about Dietary Fat," *The Journal of Nutrition*, Vol. 128, no. 2 (February 1998), pp. 449S–452S.

4 Food and Agricultural Organization (FAO) of the United Nations, *World Agriculture: Towards 2015/2030 – An FAO Perspective*, Section 3.3. Livestock Commodities, at www.fao.org/3/y4252e/y4252e0.htm.

5 My deepest thanks to Donald Worster who sent me lengthy comments on an early version of this book in a long email of June 24, 2019. He noted pointedly the importance of considering the capital-

ist essence of industrial food production, as well as the importance of considering the consumer as a major actor. Please see his *Dust Bowl* (New York: Oxford University Press, 1979) for a discussion of how technology, agricultural knowledge and the capitalist impulse came together to create the Dust Bowl. "Drought" alone as a cause of the Dust Bowl is a superficial explanation.

6 William Lippincott, *Poultry Production* (Philadelphia and New York: Lea & Febiger, 1914), pp. 20–6.
7 Lippincott, *Poultry Production*, pp. 32–4.
8 Pew Commission on Industrial Farm Animal Production, *Putting Meat on the Table: Industrial Farm Production in America* (Philadelphia: Pew, 2008), p. 5.
9 James MacDonald and William McBride, *The Transformation of Livestock Agriculture: Scale, Efficiency and Risks*, Economic Research Service Information Bulletin 43 (Washington, DC: USDA, January 2009), p. 6.
10 William Boyd, "Making Meat: Science, Technology, and American Poultry Production," *Technology and Culture*, vol. 42, no. 4 (Oct. 2001), p. 634.
11 MacDonald and McBride, *The Transformation of Livestock Agriculture*, pp. 2–3.
12 MacDonald and McBride, *The Transformation of Livestock Agriculture*, pp. 2–3.
13 MacDonald and McBride, *The Transformation of Livestock Agriculture*, p. 6.
14 MacDonald and McBride, *The Transformation of Livestock Agriculture*, p. 6.
15 Judy A. Mills and Christopher Servheen, "The Asian Trade in Bears and Bear Parts: Impacts and Conservation Recommendations," in *Bears: Their Biology and Management*, vol. IX, part 1: *A Selection of Papers from the Ninth International Conference on Bear Research and Management, Missoula, Montana, February 23–28, 1992* (1994), pp. 161–7.
16 Kaitlyn-Elizabeth Foley, Carrie Stengel and Chris Shepherd, *Pills, Powders, Vials and Flakes: The Bear Bile Trade in Asia* (Petaling Jaya, Selangor, Malaysia: TRAFFIC Southeast Asia, 2011), p. v.
17 Simon Denyer, "China's Bear Bile Industry Persists Despite

Growing Awareness of the Cruelty Involved," *Washington Post*, June 3, 2018, at www.washingtonpost.com/world/asia_pacific/from-hemorrhoids-to-hangovers-bears-bile-is-treasured-in-china-and-thats-bad-for-captive-bears/2018/06/02/fdb431da-5363-11e8-b00a-17f9fda3859b_story.html.

18 E. Livingstone, L. Gomez and J. Bouhuys, "A Review of Bear Farming and Bear Trade in Lao People's Democratic Republic," *Global Ecology and Conservation*, vol. 13 (January 2018), at www.sciencedirect.com/science/article/pii/S2351989417302615.

19 European Parliament, "Farming of Bear Bile in China," January 17, 2006, at www.europarl.europa.eu/sides/getDoc.do?pubRef=-//EP//TEXT+TA+P6-TA-2006-0008+0+DOC+XML+V0//EN&language=EN.

20 Rachel Nuwer, "Asia's Illegal Wildlife Trade Makes Tigers a Farm-to-Table Meal," *New York Times*, June 5, 2017, at www.nytimes.com/2017/06/05/science/animal-farms-southeast-asia-endangered-animals.html.

21 On the "encounter between mechanization and organic substance," see Boyd, "Making Meat," pp. 631–64.

22 Boyd, "Making Meat," pp. 632–3.

23 Deborah Fitzgerald, *The Business of Breeding: Hybrid Corn in Illinois, 1890-1940* (Ithaca: Cornell University Press, 1990).

24 Deborah Fitzgerald, *Every Farm a Factory: The Industrial Ideal in American Agriculture* (New Haven, Conn.: Yale University Press, 2003). Margaret Derry offers discussion of the impact of genetics on livestock breeding, and of breeding on modern genetics. She suggests that genetics contributed little to the practice of breeding until the second half of the twentieth century. Here, of course, big businesses began to rally round and support financially the development of the enterprise. See Margaret E. Derry, *Art and Science in Breeding: Creating Better Chickens* (Toronto and London: University of Toronto Press, 2012).

25 Page Smith and Charles Daniel, *The Chicken Book: Being an Inquiry into the Rise and Fall, Use and Abuse, Triumph and Tragedy of Gallus Domesticus* (Boston: Little, Brown, 1975). Smith and Daniel described poultry genetics from the 1930s at Kimber Farms in Fremont, California, with chickens bred for meat or eggs in breed-

ing factories called "farms" to keep the public believing that agriculture remained bucolic. Smith and Charles also explore chicken in folklore, medicine and literature, and such cultural phenomena as cock-fighting.

26 Roger Horowitz, *Putting Meat on the American Table: Taste, Technology, Transformation* (Baltimore: Johns Hopkins University Press, 2006).

27 Maryn McKenna, *Big Chicken: The Incredible Story of How Antibiotics Created Modern Agriculture and Changed the Way the World Eats* (Washington DC: National Geographic, 2017).

28 Ruth Harrison, *Animal Machines* (New York: Ballantine, 1964). See also Karen Sayer, "*Animal Machines*: The Public Response to Intensification in Great Britain, c. 1960 – c. 1973," *Agricultural History*, vol. 87, no. 4 (Fall 2013), pp. 473–501.

29 See United Poultry Concerns, "About United Poultry Concerns," October 12, 2008, at www.upc-online.org/more_about_upc.html, and Karen Davis, *Prisoned Chickens, Poisoned Eggs* (Machipongo, Va.: UPC, 2009).

30 Annie Potts, *Chicken* (London: Reaktion Books, 2012).

31 Joseph Barber, *Chicken: A Natural History* (Princeton University Press, 2017).

32 Pew Commission, *Putting Meat on the Table*, p. 23.

33 See, most importantly, Daniel Imhoff, Douglas Tompkins and Roberto Carra, *CAFO: The Tragedy of Industrial Animal Factories* (2011–13), at www.cafothebook.org/thebook.htm.

34 See Harriet Ritvo, "Animal Planet," *Environmental History*, vol. 9, no. 2 (Apr. 2004), pp. 204–20, for a survey article on the kinds of issues and questions about animals that environmental historians might ponder, and why. See also *Noble Cows and Hybrid Zebras: Essays on Animals and History* (Charlottesville: University of Virginia Press, 2010) for a collection of essays Ritvo has witten on animal–human–nature interactions.

35 On this phenomenon in the dairy industry, see Sally McMurry, "The Impact of Sanitation Reform on the Farm Landscape in U.S. Dairying, 1890–1950," *Buildings & Landscapes: Journal of the Vernacular Architecture Forum*, vol. 20, no. 2 (Fall 2013), pp. 22–47.

36 Richard Tucker, *Insatiable Appetite: The United States and the*

Ecological Destruction of the Tropical World (Berkeley: University of California Press, 2000). See also Josephson, *Industrialized Nature* (Washington, DC: Island Press, 2002) and *Fish Sticks, Sports Bras and Aluminum Cans* (Baltimore: Johns Hopkins University Press, 2010), both of which indicate that colonial, trade, and other distant agricultural and economic activities have significant impacts on local people and environments.

37 Tucker, *Insatiable Appetite*.

38 Hiram M. Drache, "Midwest Agriculture: Changing with Technology," *Agricultural History*, vol. 50, no. 1 (Jan., 1976), p. 291.

39 Drache, "Midwest Agriculture," p. 292.

40 Drache, "Midwest Agriculture," p. 299.

41 Drache, "Midwest Agriculture," p. 302.

42 Ingolf Vogeler, *The Myth of the Family Farm* (Boulder: Westview Press, 1981).

43 Adam Andrzejewski, "Mapping The U.S. Farm Subsidy $1M Club," *Forbes*, August 14, 2018, at www.forbes.com/sites/adamandrzejewski/2018/08/14/mapping-the-u-s-farm-subsidy-1-million-club.

44 Berkeley Hill, "The 'Myth' of the Family Farm: Defining the Family Farm and Assessing its Importance in the European Community," *Journal of Rural Studies*, vol. 9, no. 4 (1993), pp. 359–70.

45 MacDonald and McBride, *The Transformation of Livestock Agriculture*, p. 18.

46 Paul Crenshaw, "Thinking of Chickens," *Southwest Review*, vol. 97, no. 4 (2012), pp. 556–7.

47 Crenshaw, "Thinking of Chickens," pp. 557–9.

48 Crenshaw, "Thinking of Chickens," p. 560.

49 Zhang Min et al. "Global Genomic Diversity and Conservation Priorities for Domestic Animals Are Associated with the Economies of their Regions of Origin," *Scientific Reports*, vol. 8, no. 11677 (Aug. 3, 2018), doi:10.1038/s41598-018-30061-0, and Paul Sorene, "The Chicken of Tomorrow (1948): Mankind's Quest for a Better Hen," *Flashbak*, June 21, 2016, at https://flashbak.com/the-chicken-of-tomorrow-1948-mankinds-quest-for-a-better-hen-62821.

50 William Muir, Hans Cheng, et al., "Genome-wide Assessment of Worldwide Chicken SNP Genetic Diversity Indicates Significant Absence of Rare Alleles in Commercial Breeds," *PNAS* (*Proceedings*

of the National Academy of Sciences), vol. 105, no. 45 (Nov. 11, 2008), pp. 17312–17, doi:10.1073/pnas.0806569105.

1 Chicken Culture

1 Shakespeare, *Twelfth Night*, Act 1, Scene 3. Shakespeare's Sonnet 143 mentions a housewife running to catch one of her chickens, but this is a metaphor for his beloved running from him as he chases her from a distance. See Shakespeare, "Shakespeare's Sonnets, Sonnet CLXIII," at www.shakespeares-sonnets.com/sonnet/143.

2 Jan Dumolyn and Jelle Haemers, "'A Bad Chicken was Brooding': Subversive Speech in Late Medieval Flanders," *Past & Present*, no. 214 (February 2012), pp. 45–86.

3 Geoffrey Chaucer, "The Nun's Priest's Tale," at https://chaucer.fas.harvard.edu/pages/nuns-priests-tale.

4 Luc Hoornaert and Kris Vlegels, *Chicken on the Menu* (Tielt, Belgium: Lannoo, 2017), p. 5.

5 "A Chicken for Every Pot" political ad, October 30, 1928, at https://iowaculture.gov/history/education/educator-resources/primary-source-sets/great-depression-and-herbert-hoover/chicken.

6 George Wilson, *The Commendation of Cockes, and Cock-fighting; Wherein is shewed, that Cocke-fighting was before the coming of Christ* (1607).

7 Muhammad A. Kavesh, "From the Passions of Kings to the Pastimes of the People: Pigeon Flying, Cockfighting, and Dogfighting in South Asia," *Pakistan Journal of Historical Studies*, vol. 3, no. 1 (Summer 2018), pp. 61–83.

8 Clifford Geertz, "Deep Play: Notes on the Balinese Cockfight," in Geertz, *The Interpretation of Cultures* (New York: Basic Books, 1973).

9 "Washington Keeps Cockfighting Legal in Three States," at https://abcnews.go.com/WNT/story?id=131142&page=1.

10 Mir24.tv, "Нигерийцам запретили проносить живых кур на стадион ЧМ в Калининграде," June 13, 2018, at https://mir24.tv/news/16309476/nigeriicev-s-zhivymi-kurami-ne-pustyat-na-tribuny-chm-po-futbolu-v-kaliningrade. Apparently, Nigerian fans were refused the same request at the 2010 World Cup in South Africa. The Nigerians colored the chicken in the team's colors and

bound its claws in black tape; one Nigerian fan, John Okoro, called the Kaliningrad refusal "ridiculous."

11 Deborah E. Popper, "'Great Opportunities for the Many of Small Means': New Jersey's Agricultural Colonies," *Geographical Review*, vol. 96, no. 1 (Jan. 2006), pp. 24–49.

12 Leonard G. Robinson, "Agricultural Activities of the Jews in America," *The American Jewish Year Book*, vol. 14: *The Jew and Agriculture* (September 12, 1912, to October 1, 1913 / 5673), pp. 21–115.

13 P. H. Lawler, *The Poultry Industry of Petaluma, CAL*, Bureau of Animal Industry Circular 92 (Washington, DC: GPO, 1906), pp. 313–16.

14 Lawler, *The Poultry Industry of Petaluma*, p. 320.

15 Stuart Thompstone, "'Bab'ye Khozyaystvo': Poultry-Keeping and Its Contribution to Peasant Income in Pre-1914 Russia," *Agriculture History Review*, vol. 40, no. 1 (1992), pp. 52–63.

16 Joanna Bourke, "Women and Poultry in Ireland, 1891–1914," *Irish Historical Studies*, vol. 25, no. 99 (May 1987), p. 293.

17 As cited in Bourke, "Women and Poultry in Ireland," p. 294.

18 Bourke, "Women and Poultry in Ireland,", pp. 295–6.

19 Bourke, "Women and Poultry in Ireland," pp. 298–304.

20 Nick DiUlio, "Visionary Vineland," *New Jersey Monthly*, July 11, 2011, at https://njmonthly.com/articles/jersey-living/visionary-vineland. See also Charles Landis, *The Founder's Own Story of the Founding of Vineland, New Jersey* (Vineland Historical and Antiquarian Society, 1903).

21 Arthur Goldhaft, *The Golden Egg* (New York: Horizon Press, 1957), and Cumberland County, New Jersey, "Arthur Goldhaft: Pioneering Vet Put 'a Chicken in Every Pot,'" at www.co.cumberland.nj.us/Arthur-goldhaft.

22 "Queen Victoria's Chickens," at www.chickens.allotment-garden.org/chicken-keeping/queen-victoria-poultry-keeper/queen-victoria-chickens; "Queen Victoria Endorsement for Poultry Breeder," at www.chickens.allotment-garden.org/chicken-keeping/queen-victoria-poultry-keeper/queen-victoria-endorsement-poultry; and "Captain Bligh Gifts Queen Victoria Cochin Chicken" at http://blog.chickenwaterer.com/2015/03/cochin-chicken-amazing-backstory.html.

23 "Queen Victoria's Chickens," "Queen Victoria Endorsement for Poultry Breeder," and "Captain Bligh Gifts Queen Victoria Cochin Chicken."
24 Meg Groeling, "War Chicken," *Emerging Civil War*, February 20, 2012, at https://emergingcivilwar.com/2012/02/20/war-chicken.
25 Gustav Süs, *The Adventures of Two Chickens*, trans. from German (Philadelphia: W. P. Hazard, 1857).
26 E. Boyd Smith, *Chicken World* (New York and London: G. E. Putnam's, 1910). Smith, born in St. John, New Brunswick, and raised in Boston, studied painting abroad before turning to children's picture books as illustrator and storyteller.
27 Lucy Sprague Mitchell, "The Rooster and the Hens," in *Here and Now Story Book*, illus. Hendrik Willem Van Loon (New York: E. P. Dutton and Co., 1921). Mitchell attended Radcliffe College, but overcame segregation of males from females in the sciences to work in the Museum of Comparative Zoology and its Radcliffe Zoological Laboratory.
28 Walter de la Mare, *Peacock Pie: A Book of Rhymes* (London: Constable and Company Ltd., 1920), p. 13.
29 USDA Library, "Frost on Chickens," www.nal.usda.gov/exhibits/ipd/frostonchickens.
30 Robert Frost, "The Cockerel Buying Habit," *Farm Poultry*, February 1, 1904, p. 54.
31 Frost, "A Blue Ribbon at Amesbury," *Atlantic Monthly*, vol. 157, no. 4 (April 1936).
32 Robert Terry, Jr., "Development and Evolution of Agriculturally Related Merit Badges Offered by the Boy Scouts of America," *Journal of Agricultural Education*, vol. 54, no. 2 (2013), pp. 70–84. See p. 75 for the first poultry merit page.
33 Boy Scouts of America, *Boy Scouts Handbook*, 1st edition (Red Bank, NJ: Boy Scouts, 1911).
34 Anon., "7,000 Birds Are Entered; Poultry Show Opens at Madison Square Garden Tuesday," *New York Times*, January 1, 1919, at www.nytimes.com/1919/01/19/archives/7000-birds-are-entered-poultry-show-opens-at-madison-square-garden.html.
35 Alexis Coe, "Today We're Eating the Winners of the 1948 Chicken of Tomorrow Contest," *Modern Farmer*, May 12, 2014, at https://

modernfarmer.com/2014/05/today-eating-winners-1948-chicken-tomorrow-contest.

36 Anon., "Poultry Extension Work in Georgia," *Poultry Science*, vol. 2, no. 4 (January 1, 1916), p. 32.

37 University of California, Agriculture and Natural Resources, "4-H History," at http://4h.ucanr.edu/About/History. See also Sarah McColl, "4-H: Indoctrination Nation," *Modern Farmer*, July 25, 2017, at https://modernfarmer.com/2017/07/4-h-indoctrination-nation. See also NCState, "Race in the Extension Service," at https://nceats.omeka.chass.ncsu.edu/exhibits/show/raceandgender/racial_segregation/race_extension_service.

38 USDA, *Poultry in Family Meals. A Guide for Consumers*, Home and Garden Bulletin 110 (Washington, DC: USGPO, 1966).

39 R. V. Hicks and Harry R. Lewis, "With the Signing of World Peace the National War Emergency Poultry Federation Ceases its Existence," *Poultry Science*, vol. 2, no. 7 (April 1, 1919), pp. 55–6.

40 USDA, Production and Marketing Administration, Information Service, Western Area, radio script, "Food Fights for Freedom – At Home and Abroad," Weekly Script no. 100, 1946. The US helped Czechoslovakia re-establish its poultry industry with 30 crates of hatching eggs – 400,000 eggs – using an air shipment to get them to incubators.

41 Alexis Coe, "Today We're Eating the Winners of the 1948 Chicken of Tomorrow Contest," *Modern Farmer*, May 12, 2014, at https://modernfarmer.com/2014/05/today-eating-winners-1948-chicken-tomorrow-contest. See also Horowitz, *Putting Meat on the American Table*, pp. 126-27, and "Chicken of Tomorrow," film, 1948 at https://archive.org/details/Chickeno1948.

42 H. L. Shrader, "The Chicken-of-Tomorrow Program: Its Influence on 'Meat-Type' Poultry Production," *Poultry Science*, vol. 31, no. 1 (January 1952), pp. 3–11; and Coe, "Today We're Eating the Winners."

43 On these Arbor Acre birds today, see http://ap.aviagen.com/brands/arbor-acres/products/arbor-acres-plus.

44 Coe, "Today We're Eating the Winners."

45 William Muir et al., "Genome-wide Assessment of Worldwide Chicken SNP Genetic Diversity Indicates Significant Absence

of Rare Alleles in Commercial Breeds," *PNAS* (*Proceedings of the National Academy of Sciences*), vol. 105, no. 45 (November 11, 2008), pp. 17312–17.

46 "State Poultry Shows" at www.poultryshowcentral.com/State.html; Trade Fair Dates, "Deutsche Junggeflügelschau Hanover" at www.tradefairdates.com/Deutsche-Junggefluegelschau-M1249/Hanover.html; and Events Eye, "Livestock & Poultry Trade Shows in Indonesia (Asia – Pacific)," at www.eventseye.com/fairs/zst1_trade-shows_asia-pacific_livestock-poultry.html#ac136.

47 Brazilian Chicken, "Brazilian Poultry Farming," at www.brazilianchicken.com.br/en/poultry-industry/poultry-farming.

48 NOVOGEN, "Festa do Ovo 2018, Bastos," www.novogen-layer.com/newsroom/novogen-press-releases/498-festa-do-ovo-2018-bastos.html; and Globos, "Festa do Ovo comemora os 88 anos da cidade de Bastos," May 7, 2016, at http://redeglobo.globo.com/sp/tvtem/noticia/2016/07/festa-do-ovo-comemora-os-88-anos-da-cidade-de-bastos.html.

49 Holly Richmond, "These Beauty Pageant Contestants Are Chickens," *Grist*, Mar. 27, 2014, at https://grist.org/living/these-beauty-pageant-contestants-are-chickens. See also "Malaysian Ayam Serama Chicken: The Smallest Chicken in the World," at www.raising-chickens.org/Serama-Chicken.html.

50 N. V. Duc and T. Long, *Poultry Production Systems in Vietnam*, GCP/RAS/228/GER Working Paper 4 (Rome: FAO, 2008). It should be noted that Avian flu has been a significant problem in Vietnam over the last twenty years, with massive breakouts beginning in 2003 that killed 12 people and resulted in the culling of millions of birds in 2004–5, and with outbreaks on a lesser scale continuing to the present. In 2018 HPAI hit Vietnam and the Netherlands very hard.

51 Ngoc Thanh, "Mister and Miss Chicken: Inside the Poultry Beauty Contest in Vietnam," *VN Express*, January 4, 2017, at https://e.vnexpress.net/news/travel-life/mister-and-miss-chicken-inside-the-poultry-beauty-contest-in-vietnam-3523276.html. On Vietnam, see also "Vietnamese Poultry, Livestock Industries Unprepared for Integration, Face Deep Trouble," Tuoitrenews.vn, July 7, 2015, at http://tuoitrenews.vn/business/29103/vietnamese-

poultry-livestock-industries-unprepared-for-integration-face-deep-trouble.

52 Arab News, "Gulf Farmers to Attend 'Miss Chicken' Pageant in KSA," October 2, 2013, at www.arabnews.com/news/466414.

53 William Haun, "Power in the Blood: Animal Sacrifice in West Africa," *International Mission Board*, May 7, 2018, at www.imb.org/2018/05/07/animal-sacrifice.

54 Sandra Klopper, "Initiation of Healers and Ritual Specialists: Ceremonies of South Africa," *South Africa Online*, 2018, at http://southafrica.co.za/initiation-of-healers-and-ritual-specialists.html. See also K. C. MacDonald, "Why Chickens? The Centrality of the Domestic Fowl in West African Ritual and Magic," in K. Ryan and P. J. Crabtree, eds., *The Symbolic Role of Animals in Archaeology* (Philadelphia: Museum Applied Science Center for Archaeology, 1995), pp. 50–6.

55 *The Alliance to End Chickens as Kaporos, et al., Plaintiffs–Appellants* v. *The New York City Police Department, et al., Defendants–Respondents, Congregation Beis Kosov Miriam Lanynski, et al., Defendants*, Decided: May 6, 2017, at https://caselaw.findlaw.com/ny-supreme-court-appellate-division/1863344.html.

56 Mara Miele, "Killing Animals for Food: How Science, Religion and Technologies Affect the Public Debate About Religious Slaughter," *Food Ethics*, vol. 1, no. 1 (June 2016), pp. 47–60.

2 Ecology and Industry

1 Alfred Crosby, *Ecological Imperialism* (Cambridge University Press, 1986).

2 Lee Perry-Gal et al., "Earliest Economic Exploitation of Chicken Outside East Asia: Evidence from the Hellenistic Southern Levant," *PNAS*, vol. 112, no. 32 (August 11, 2015), pp. 9849–54.

3 Joris Peters et al., "Holocene Cultural History of Red Jungle Fowl (*Gallus Gallus*) and its Domestic Descendant in East Asia," *Quaternary Science Reviews*, vol. 142 (June 2016), pp. 102–19.

4 Jacqueline Pitt et al., "New Perspectives on the Ecology of Early Domestic Fowl: An Interdisciplinary Approach," *Journal of Archaeological Science*, vol. 74 (2016), pp. 1–10. See also Carys E. Bennett, Richard Thomas et al., "The Broiler Chicken as a Signal

of a Human Reconfigured Biosphere," December 12, 2018, Royal Society Open Science, at https://doi.org/10.1098/rsos.180325, and Richard Thomas, Matilda Holmes and James Morris, "'So Bigge as Bigge May Be': Tracking Size and Shape Change in Domestic Livestock in London (AD 1220–1900)," *Journal of Archaeological Science*, vol. 40, no. 8 (2013), pp. 3309–25. On the project on the Cultural and Scientific Perception of Human–Chicken Interaction, see www.scicultchickens.org.

5 Magda Mehdawy and Amr Hussein, *The Pharaoh's Kitchen* (Cairo and New York: The American University in Cairo Press, 2010), pp. 61–5.

6 J. M. Mwacharo, G. Bjørnstad, J. L. Han et al., "The History of African Village Chickens," *African Archaeological Review*, vol. 30, no. 1 (2013), pp. 97–114, p. 99.

7 Alice Storey et al., "Polynesian Chickens in the New World: A Detailed Application of a Commensal Approach," *Archaeology in Oceania*, vol. 48, no. 2 (July 2013), pp. 101–19. See also S. Castello, "Existence of Chickens in America Before its Discovery and Conquest," in "Proceedings of the Seventh World Poultry Congress and Exposition, Cleveland, Ohio, USA, 28 July to 7 August 1939," pp. 459–62.

8 Mwacharo et al., "The History of African Village Chickens."

9 Mwacharo et al., "The History of African Village Chickens," p. 98.

10 Robyn Alders and Robert Pym, "Village Poultry: Still Important to Millions, Eight Thousand Years After Domestication," *World's Poultry Science Journal*, vol. 65, no. 2 (2009), pp 181–2.

11 Thorleif Schjelderup-Ebbe, "Social Behavior of Birds," in *A Handbook of Social Psychology* (Worcester, Mass.: Clark University Press, 1935), pp. 947–72.

12 Horace Atwood, *Poultry on the Farm* (Morgantown: WVU Agricultural Experiment Station, 1915), p. 3.

13 Atwood, *Poultry on the Farm*, p. 6.

14 W. P. Wheeler, *Feeding Experiments with Laying Hens: Efficiency of Whole and Ground Grains*, Bulletin 106 (Geneva: New York Agricultural Experiment Station, August, 1896).

15 J. H. Stewart and Horace Atwood, *Poultry Experiments*, Bulletin 71

(Morgantown: West Virginia University Agricultural Experiment Station, December 1900).

16 W. S. Jacobs, *Farm Poultry*, Bulletin 99 (Fayetteville, Ark.: Arkansas Agricultural Experiment Station, 1908), pp. 150–1.

17 Jacobs, *Farm Poultry*, p. 169.

18 Rob R. Slocum, *The Selection and Care of Poultry Breeding Stock* (Washington, DC: USDA, 1920), p. 5.

19 Slocum, *The Selection and Care of Poultry Breeding Stock*, pp. 8, 10.

20 R. A. Kock, "Breeding Centres for Poultry in Denmark," *Poultry Science*, volume 2-1, no. 9 (July 1915), pp. 70–1.

21 Edith Bradley and Bertha La Motte, *The Lighter Branches of Agriculture*, vol. VI (London: Chapman and Hall, 1903), pp. 199–200.

22 Oscar Erf, *The Hen's Place on the Farm*, Bulletin 150 (Manhattan, Kans.: Kansas State Agricultural Experiment Station, October 1907), p. 47.

23 Erf, *The Hen's Place*, p. 41.

24 Erf, *The Hen's Place*, p. 47.

25 Erf, *The Hen's Place*, p. 50.

26 Harry M. Lamon and Jos. Wm Kinghorne, "Illustrated Poultry Primer," in *Farmers' Bulletins Nos. 1026–1050* (Washington, DC: US Dept. of Agriculture, 1919), pp. 7, 9, 14–15.

27 *Poultry Herald* (September 1920), p. 284.

28 Milo Hastings, *Physical Culture Food Directory* (New York: MacFadden Publications, 1927).

29 Milo Hastings, "A Million Chicks to the Acre," *Scientific American*, vol. 113, no. 12 (September 18, 1915), pp. 247, 257–60.

30 Gary M. Landsberg, "Social Behavior of Chickens," in *Merck Veterinary Manual*, at www.merckvetmanual.com/behavior/normal-social-behavior-and-behavioral-problems-of-domestic-animals/social-behavior-of-chickens.

31 J. Mench and L. J. Keeling, "The Social Behaviour of Domestic Birds," in L. J. Keeling and H. W. Gonyou, eds., *Social Behaviour in Farm Animals* (Wallingford, UK: CABI Publishing, 2001), pp. 177–209; B. O. Hughes and D. G. M. Wood-Gush, "Agonistic Behaviour in Domestic Hens: The Influence of Housing Method and Group Size," *Animal Behaviour*, vol. 25 (1977), pp. 1056–62; J. M. Mauldin, "Applications of Behaviour to Poultry Management,

Symposium: Quantifying the Behaviour of Poultry," *Poultry Science*, vol. 71 (1992), pp. 634–42; G. McBride and F. Foenander, "Territorial Behaviour in Flocks of Domestic Fowls," *Nature*, vol. 194 (1962), p. 102; G. McBride, J. W. James and R. N. Shoffner, "Social Forces Determining Spacing and Head Orientation in a Flock of Domestic Hens," *Nature*, vol. 197 (1963), pp. 1272–3; and P. B. Siegel, "The Role of Behaviour in Poultry Production: A Review of Research," *Applied Animal Ethology*, vol. 11 (1984), pp. 299–316.

32 Mench and Keeling, "The Social Behaviour of Domestic Birds."

33 M. A. Jull, *Farm Poultry Raising*, Farmers' Bulletin 1524 (Washington, DC: USDA, 1927), p. 10.

34 E. F. Armstrong, "Chickens," *Journal of the Royal Society of Arts*, vol. 87, no. 4510 (April 28, 1939), pp. 626–8.

35 J. V. Craig, "Aggressive Behavior of Chickens: Some Effects of Social and Physical Environments," paper presented at the 27th Annual National Breeder's Roundtable, Kansas City, Missouri, May 11, 1978, at https://pdfs.semanticscholar.org/13f6/8d1b90c892ee8be4e2f3aaca50bab0759d4d.pdf.

36 William H. Williams, *Delmarva's Chicken Industry: 75 Years of Progress* (Georgetown, Del.: Delmarva Poultry Industry, Inc., 1998), pp. ix–x.

37 Williams, *Delmarva's Chicken Industry*, pp. 17–21, 22–4, 31, 44–5.

38 Arthur Gannon, "Georgia's Broiler Industry," *The Georgia Review*, vol. 6, no. 3 (Fall 1952), p. 311.

39 Gannon, "Georgia's Broiler Industry," pp. 306–10.

40 Carl Weinberg, "Big Dixie Chicken Goes Global: Exports and the Rise of the North Georgia Poultry Industry," *Business and Economic History. On-Line*, vol. 1 (2003), at https://thebhc.org/sites/default/files/Weinberg_0.pdf, and "Jesse Jewell (1902–1975)," *New Georgia Encyclopedia*, January 14, 2005, at www.georgiaencyclopedia.org/articles/business-economy/jesse-jewell-1902-1975.

41 Gannon, "Georgia's Broiler Industry," p. 311.

42 Gannon, "Georgia's Broiler Industry," pp. 312–15.

43 Verel W. Benson and Thomas J. Witzig, *The Chicken Broiler Industry: Structure, Practices and Costs*, Agricultural Economic Report 381

(Washington, DC: USDA, 1971), pp. i–ii. NB: The publication date on the brochure is 1971, but the information in it indicates a typographical error, and it should be 1976.

44 Benson and Witzig, *The Chicken Broiler Industry*, pp. 1–15.

45 US International Trade Commission, *Industry and Trade Summary: Poultry*, Publication 2520 (AG-6) (Washington, DC: US ITC, June 1992), p. 4.

46 US International Trade Commission, *Industry and Trade Summary*, p. 5.

47 H. W. Cheng, "Breeding of Tomorrow's Chickens to Improve Well-being," *Poultry Science*, vol. 89, no. 4 (April 1, 2010), pp. 805–13.

48 Cheng, "Breeding of Tomorrow's Chickens."

49 Lori Marino, "Thinking Chickens: A Review of Cognition, Emotion, and Behavior in the Domestic Chicken," *Animal Cognition*, vol. 20, no. 2 (2017), pp. 127–47.

50 Marino, "Thinking Chickens."

51 Marino, "Thinking Chickens."

52 Mench and. Keeling, "The Social Behaviour of Domestic Birds."

53 Marino, "Thinking Chickens."

54 Mench and Keeling, "The Social Behaviour of Domestic Birds."

55 I. Hansen and B. O. Braastad, "Effect of Rearing Density on Pecking Behaviour and Plumage Condition of Laying Hens in Two Types of Aviary," *Applied Animal Behavioural Science*, vol. 40 (1994), pp. 263–72.

56 M. Dawkins, "Environmental Preference Studies in the Hen," *Animal Regulation Studies*, vol. 3 (1980), pp. 57–63. This journal has been published by the UN FAO since 1977.

57 M. S. Dawkins and R. Layton, "Breeding for Better Welfare," *Animal Welfare*, vol. 21 (2012), pp. 147–55.

58 Timothy Pachirat, *Every Twelve Seconds: Industrialized Slaughter and the Politics of Sight* (New Haven, Conn.: Yale University Press, 2011), pp. 3–4. In the massive slaughterhouse where Pachirat worked, 2,400 cows were slaughtered daily. These were no longer anything except homogenized raw material (p. 40).

59 Dawkins and Layton, "Breeding for Better Welfare," p. 147. See also M. S. Dawkins, *Why Animals Matter: Animal Consciousness,*

Animal Welfare and Human Well-Being (Oxford University Press, 2012), and Dawkins and R. Bonney, *The Future of Animal Farming: Renewing the Ancient Contract* (Blackwell: Oxford, 2007).

60 Felicity Lawrence, "If Consumers Knew How Farmed Chickens Were Raised, They Might Never Eat Their Meat Again," *The Guardian*, April 24, 2016, at www.theguardian.com/environment/2016/apr/24/real-cost-of-roast-chicken-animal-welfare-farms.

61 Claire Andreasen, Anna Spickler and Bernt Jones, "Swedish Animal Welfare Regulations and Their Impact on Food Animal Production," *Journal of the American Veterinary Medical Association*, vol. 227, no. 1 (July 1, 2005), pp. 34–40.

62 Coline Brame, "Volaille de chair: contrôles officiels du bien-être, l'affaire de tous!" *Proagri* (Chambres d'Agriculture Bretagne), May 20, 2016, pp. 32–3, at www.bretagne.synagri.com/ca1/PJ.nsf/TECHPJPARCLEF/27068/$File/Aviculture-volaille-de-chair-contr%C3%B4les-bien-%C3%AAtre-affaire-de-tous2016-05.pdf?OpenElement.

63 A. Devos, "Rupture of the Gastrocnemius Tendon in Chickens," *Avian Diseases*, vol. 7, no. 4 (Nov. 1963), pp. 451–6.

64 Sipke Joost Hiemstra and Jan Ten Napel, *Study of the Impact of Genetic Selection on the Welfare of Chickens Bred and Kept for Meat Production*, Final Report, IBF International Consulting, January 2013, pp. 10–11.

65 Dawkins and Layton, "Breeding for Better Welfare."

66 The concerns have a legal foundation: the need to meet the requirements of EFSA, "Scientific Opinion on the Influence of Genetic Parameters on the Welfare and the Resistance to Stress of Commercial Broilers," *EFSA Journal*, 8 (2010), 1666; and EFSA, "Scientific Opinion on Welfare Aspects of the Management and Housing of the Grand-Parent and Parent Stocks Raised and Kept for Breeding Purposes," *EFSA Journal*, 8 (2010), 1667.

67 Hiemstra and Napel, *Study of the Impact of Genetic Selection*, pp. 7–8.

68 Hiemstra and Napel, *Study of the Impact of Genetic Selection*, pp. 7–8.

69 Dina Spector, "Ever Wondered Why Europeans Don't Refrigerate Eggs, but Americans Do?" *Business Insider*, December 1, 2014, at www.businessinsider.com/why-europeans-dont-refrigerate-eggs-2014-12.

70 On the environmental challenges of safe egg production, see Joe Fassler, "Timeline of Shame: Decades of DeCoster Egg Factory Violations," *The Atlantic*, September 16, 2010, at www.theatlantic.com/health/archive/2010/09/timeline-of-shame-decades-of-decoster-egg-factory-violations/63059.
71 Cobb Creek Farm, "Meet our Chickens," at www.cobbcreekfarm.com/pastured-poultry.

3 Chicken as Machine

1 Julien Offray de la Mettrie, *Man, A Machine* (Chicago, Ill.: Open Court, 1912), p. 1.
2 Mench and Keeling, "The Social Behaviour of Domestic Birds."
3 Sam White, "From Globalized Pig Breeds to Capitalist Pigs: A Study in Animal Cultures and Evolutionary History," *Environmental History*, vol. 16, no. 1 (Jan. 2011), p. 94.
4 John Bird, "Our Ingenious Farmers," *Challenge*, vol. 1, no. 2 (November 1952), p. 51.
5 Compassion in World Farming, "The Life of: Broiler Chickens," CWF, January 5, 2013, at www.ciwf.org.uk/media/5235306/The-life-of-Broiler-chickens.pdf.
6 Paul Crenshaw, "Thinking of Chickens," *Southwest Review*, vol. 97, no. 4 (2012), pp. 562–3.
7 William Boyd, "Making Meat: Science, Technology, and American Poultry Production," *Technology and Culture*, vol. 42, no. 4 (Oct. 2001), p. 662.
8 Homer Jackson, "Animal Food for Poultry," *Poultry Science*, vol. 2, no. 7 (May 1, 1915), pp. 53–4.
9 James MacDonald and William McBride, *The Transformation of Livestock Agriculture: Scale, Efficiency and Risks*, Economic Research Service Information Bulletin 43 (Washington, DC: USDA, January 2009), p. 23.
10 See Alan Marcus, *Cancer from Beef: DES, Federal Food Regulation, and Consumer Confidence* (Baltimore: Johns Hopkins University Press, 1994), for a discussion of the 20-year battle to ban DES (diethylstilbestrol, a known carcinogen, yet capable of accelerating the fattening of beef by up to 10 percent and getting them to market a month earlier), and generally to understand the food-safety regulatory process.

11 National Chicken Council, "Genetically Modified Organism (GMO) Use in the Chicken Industry," June 5, 2013, at www.nationalchickencouncil.org/genetically-modified-organism-gmo-use-in-the-chicken-industry.
12 Non-GMO Project, "Animal Feed is the Key to a Non-GMO Future," October 19, 2017, at www.nongmoproject.org/blog/animal-feed-is-the-key-to-a-non-gmo-future.
13 Non-GMO Project, "Animal Feed is the Key to a Non-GMO Future."
14 According to one study, some lighting programs have a central purpose of slowing the early growth rate of broilers, thus allowing birds to achieve physiological maturity before maximal rates of muscle mass accretion. See H. A. Olanrewaiu et al., "A Review of Lighting Programs for Broiler Production," *International Journal of Plant Science*, vol. 5, no. 4 (2006), p. 301. See Hy-Line, "Basic Rules for Lighting," at https://www.hyline.com/aspx/redbook/redbook.aspx?s=4&p=23, for a discussion of lighting requirements for genetically engineered broilers over the duration of their short lives.
15 J. A. Renden, E. T. Moran, Jr. and S. A. Kincade, "Lighting Programs for Broilers that Reduce Leg Problems Without Loss of Performance or Yield Poultry," *Science*, vol. 75, no. 11 (1996), pp. 1345–50.
16 A. Deep et al., "Effect of Light Intensity on Broiler Production, Processing Characteristics, and Welfare," *Poultry Science*, vol. 89, no. 11 (2010), pp. 2326–33.
17 Y. Yang et al., "Artificial Light and Biological Responses of Broiler Chickens: Dose-Response," *Journal of Animal Science*, vol. 96, no. 1 (Feb. 2018), pp. 98–107.
18 Bob Alphin, "Impact of Light on Poultry," PowerPoint presentation, Department of Animal & Food Sciences Newark, Delaware, USA, at https://extension.umd.edu/sites/extension.umd.edu/files/_images/programs/poultry/Alphin%20Light%20Impact%20on%20Poultry%203-11-14.pdf.
19 Poultry Hub, "Beak Trimming," at www.poultryhub.org/health/health-management/beak-trimming, with bibliography.
20 T. G. Petrolli et al., "Effects of Laser Beak Trimming on the

Development of Brown Layer Pullets," *Brazilian Journal of Poultry Science*, vol. 19, no. 1 (Jan.–Mar. 2017), p. 123.

21 D. C. Kennard, *Chicken Vices*, Bulletin 184, Ohio Agricultural Experiment Station, vol. 22 (1937), pp. 33–9.

22 "Hand Held Beak Trimmer," www.freepatentsonline.com/3593714.pdf.

23 UPC (United Poultry Concerns), "Supersizing the Chicken," February 19, 2015, at www.upc-online.org/industry/150219_super-sizing_the_chicken.html.

24 Gary M. Landsberg, "Social Behavior of Chickens," in *Merck Veterinary Manual*, at www.merckvetmanual.com/behavior/normal-social-behavior-and-behavioral-problems-of-domestic-animals/social-behavior-of-chickens.

25 Petrolli et al., "Effects of Laser Beak Trimming," p. 123.

26 Anon., "Entrepreneur Wants a Lens in Every Chicken," *Harvard Crimson*, November 27, 1989, at www.thecrimson.com/article/1989/11/27/entrepreneur-wants-a-lens-in-every.

27 Irvin Wise and Lester Hall, Vision Control, Inc., Santa Rosa, Calif., "Distorting Contact Lenses for Fowl Patented," US Patent #3418978, December 31, 1968.

28 M. A. Jull, *Farm Poultry Raising*, Farmers' Bulletin no. 1524 (Washington, DC: USDA, 1927), pp. 24–5.

29 A. O'Connor et al., *Preparatory Work for Future Development of Four Scientific Opinions on Monitoring Procedures at Slaughterhouses*, EFSA Supporting Publications, vol. 10, no. 12 (December 2013), at https://doi.org/10.2903/sp.efsa.2013.EN-467.

30 Max Zimmerman, *The Super Market* (New York: McGraw-Hill, 1955); Ralph Cassady, *Competition and Price Making in Food Retailing* (New York: Ronald Press Company, 1962); and Edward Brand, *Modern Supermarket Operation* (New York: Fairchild Industries, 1963).

31 Harvey Levenstein, *Paradox of Plenty: A Social History of Eating in Modern America* (New York: Oxford University Press, 1993), 113–14.

32 Cory Bernat, "Supermarket Packaging: The Shift from Glass to Aluminum to Plastic," *The Atlantic*, January 25, 2012, at www.theatlantic.com/health/archive/2012/01/supermarket-packaging-the-shift-from-glass-to-aluminum-to-plastic/251875.

33 But in 1955 the US Supreme Court determined that Du Pont did not monopolize interstate commerce in violation of section 2 of the Sherman Act. The Court determined that there was not an invidious attempt to control price or exclude competition. *United States* v. *E. I. du Pont de Nemours & Co.*, 351 U.S. 377 (1956). See Donald Turner, "Antitrust Policy and the Cellophane Case," *Harvard Law Review*, vol. 70, no. 2 (Dec. 1956), pp. 281–318.

34 Bernat, "Supermarket Packaging."

35 Ziaul Ahmed and Mark Sieling, "Two Decades of Productivity Growth in Poultry Dressing and Processing," *Monthly Labor Review* (April 1987), pp. 34–9.

36 Michael Martinez, S. Anders and W. Wismer, "Consumer Preferences and Willingness to Pay for Value-Added Chicken Product Attributes," *Journal of Food Sciences*, vol. 76, no. 8 (October 1976), pp. 469–77.

37 John-Bryan Hopkins, "A History of Chicken Nuggets," March 15, 2012, at http://foodimentary.com/2012/03/15/history-chicken-nuggets.

38 Baker enjoyed food and cooking. See Robert Baker, *Barbecued Chicken and Other Meats* (Ithaca, NY: Cornell Cooperative Extension, 1950). See Robert Baker papers, 1934–97, Collection Number: 21-26-4030, Division of Rare and Manuscript Collections, Cornell University Library.

39 National Chicken Council, "What's Really in that Chicken Nugget?" 2012, at www.nationalchickencouncil.org/whats-in-those-chicken-nuggets.

40 Laura Stampler, "Usain Bolt Ate 100 Chicken McNuggets a Day in Beijing and Somehow Won Three Gold Medals," *Time*, November 4, 2013, at http://time.com/3912896/usain-bolt-chicken-mcnuggets-olympics.

41 Richard deShazo, Steven Bigler and Leigh Skipworth, "The Autopsy of Chicken Nuggets Reads 'Chicken Little,'" *American Journal of Medicine*, vol. 126 (2013), pp. 1018–19.

42 Another frequent approach is to suggest that new technologies – radiation-sterilization of food among them – will keep us and our chickens and other foods bacteria-free.

43 Laura Brown et al., "Frequency of Inadequate Chicken Cross-Contamination Prevention and Cooking Practices in Restaurants," *Journal of Food Protection*, vol. 76, no. 12 (2013), pp. 2141–5.
44 Joe W. Atkinson, "Trends in Poultry Hygiene," *Public Health Reports*, vol. 72, no. 11 (Nov. 1957), pp. 949–56.
45 Atkinson, "Trends in Poultry Hygiene."
46 Simon Dawson, "Chlorine-washed Chicken Q&A: Food Safety Expert Explains Why US Poultry is Banned in the EU," *The Conversation*, August 2, 2017, at https://theconversation.com/chlorine-washed-chicken-qanda-food-safety-expert-explains-why-us-poultry-is-banned-in-the-eu-81921.
47 J. E. Watt, "Salmonella Infection in Poultry," *Canadian Journal of Public Health*, vol. 57, no. 11 (November 1966), pp. 522–4.
48 Boyd, "Making Meat," p. 636, and OTA, *Drugs in Livestock Feed*, NTIS order #PB-298450 (Washington: OTA, June 1979), p. 3.
49 M. J. Martin, S. E. Thottathil and T. B. Newman, "Antibiotics Overuse in Animal Agriculture: A Call to Action for Health Care Providers," *American Journal of Public Health*, vol. 105, no. 12 (2015), pp. 2409-10.
50 MacDonald and McBride, *The Transformation of Livestock Agriculture*, pp. 32–3.
51 Food and Drug Administration, Center for Veterinary Medicine, *2016 SUMMARY REPORT on Antimicrobials Sold or Distributed for Use in Food-Producing Animals* (Washington, DC: FDA, 2017).
52 "H.R. 1150 (113th): Preservation of Antibiotics for Medical Treatment Act of 2013," Govtrack at www.govtrack.us/congress/bills/113/hr1150.
53 Martin et al., "Antibiotics Overuse," pp. 2409–10.
54 Nataliya Roth et al., "The Application of Antibiotics in Broiler Production and the Resulting Antibiotic Resistance in *Escherichia coli*: A Global Overview," *Poultry Science*, December 13, 2018, at https://www.ncbi.nlm.nih.gov/pmc/articles/pmc6414035.
55 Martin et al., "Antibiotics Overuse," pp. 2409–10.
56 Maryn McKenna, *Big Chicken: The Incredible Story of How Antibiotics Created Modern Agriculture and Changed the Way the World Eats* (Washington, DC: National Geographic, 2017).
57 McKenna, *Big Chicken*.

58 A. A. Saleha, “Isolation and Characterization of *Campylobacter jejuni* from Broiler Chickens in Malaysia,” *International Journal of Poultry Science*, vol. 1, no. 4 (2002), pp. 94–7.
59 O. Sahin et al., “Campylobacter in Poultry: Ecology and Potential Interventions,” *Avian Diseases*, vol. 59, no. 2 (June 2015), p. 185.
60 Sahin et al., “Campylobacter in Poultry,” p. 192.
61 European Food Safety Authority, “Foodborne Zoonotic Diseases,” www.efsa.europa.eu/en/topics/topic/foodborne-zoonotic-diseases.
62 B. Lupiani and S. M. Reddy, “The History of Avian Influenza,” *Comparative Immunology, Microbiology and Infectious Diseases*, vol. 32, no. 4 (July 2009), pp. 311–23. See also E. Perroncito “Epizoozia tifoide nei gallinacei,” *Annali di Accademia di Agricultura Torino*, vol. 21 (1878), pp. 87–126; and E. L. Stubs, “Fowl Pest,” in H. E. Biester and L. Devries, eds., *Diseases of Poultry* (Ames: Iowa State College Press, 1943), pp. 493–502.
63 W. Schäfer, “Vergleichender sero-immunologische Untersuchungen über die Viren der Influenza und klassischen Geflügelpest,” *Zeitschrift Naturforschung*, vol. 10b (1955), pp. 81–91.
64 First International Symposium on Avian Influenza, “Proceedings of the First International Symposium on Avian Influenza, Beltsville, MD, 1981,” *Avian Diseases*, vol. 47 (Special Issue) (2003).
65 On Curtice, see USDA, “Cooper Curtice,” at www.nal.usda.gov/exhibits/speccoll/exhibits/show/parasitic-diseases-with-econom/item/8292; E. F. Pernot, *Investigations of Disease of Poultry*, Bulletin 64 (Corvallis: Oregon Agricultural Experiment Station, 1900).
66 Philip B. Hadley and M. F. Kirkpatrick, “The Present Status of Investigation of the Problems of Poultry Culture,” *Poultry Science*, vol. 1, no. 1 (January 1908), pp. 59–61.
67 George Byron Morse, “Profitable Lines of Investigation in Poultry Diseases,” *Poultry Science*, vol. 1, no. 1 (January 1, 1908), p. 62.
68 Horace Atwood, “The Field of Research in Poultry Husbandry,” *Poultry Science*, vol. 1, no. 1 (January 1, 1908), p. 27.
69 W. R. Hinshaw, “The History of Avian Medicine in the United States. VII. Developments in Avian Pathology with Emphasis on Avian Practice,” *Avian Diseases*, vol. 26, no. 3 (July–Sept. 1982), pp. 462–72.
70 Atkinson, “Trends in Poultry Hygiene,” pp. 949–56.

71 Elizabeth Wilson, Milton Foter and Keith Lewis, "Public Health Aspects of Food Poisoning," *Journal of Milk and Food Technology*, vol. 20, no. 3 (March 1957), pp. 65-71.
72 James Lieberman, "Progress in Poultry Inspection and Sanitation," *Public Health Reports*, vol. 69, no. 2 (Feb. 1954), pp. 136–40.
73 Lieberman, "Progress in Poultry Inspection," pp. 136–40.
74 National Research Council, Committee on Public Health Risk Assessment of Poultry Inspection Programs, *Poultry Inspection: The Basis for a Risk-Assessment Approach* (Washington, DC: National Academy Press, 1987), p. 1.
75 National Research Council, *Poultry Inspection*, p. 2.
76 National Research Council, *Poultry Inspection*, pp. 3–4.
77 FSIS, "FSIS History," at www.fsis.usda.gov/wps/portal/informat ional/aboutfsis/history.
78 USDA, "USDA Announces Additional Food Safety Requirements, New Inspection System for Poultry Products," Press Release No. 0163.14, July 31, 2014, at www.usda.gov/media/press-releases/ 2014/07/31/usda-announces-additional-food-safety-requirements-new-inspection.
79 Food & Water Watch, "Privatized Inspection System Produces More Contaminated Chicken," January 30, 2018, www.foodandwat erwatch.org/news/privatized-inspection-system-produces-more-co ntaminated-chicken.
80 FSIS, "FSIS in Arkansas," at www.fsis.usda.gov/wps/wcm/conn ect/20931a6c-c421-42a1-8249-d8ce3eaa7cd8/AR.pdf?MOD=AJP ERES.
81 Alessandra Nicita, *Avian Influenza and the Poultry Trade*, Policy Research Working Paper 4551 (World Bank: Development Research Group, March 2008); WHO, "Avian Influenza A (H5N1) in Humans and Poultry in Viet Nam," January 13, 2004, at www.who. int/csr/don/2004_01_13/en; CNN, "S. Korea: Bird Flu Spreading," December 20, 2003, at www.cnn.com/2003/WORLD/asiapcf/east/ 12/20/skorea.bird.flu.reut; Kang Shinhye, "South Korean Poultry Farm Has Suspected Bird Flu Case," Reuters, January 20, 2007, at www.reuters.com/article/us-birdflu-korea-s/south-korean-poult ry-farm-has-suspected-bird-flu-case-idUSSEO18546520061123.
82 D. J. Alexander and I. H. Brown, "History of Highly Pathogenic

Avian Influenza," *Revue Scientifique et Technique*, vol. 28, no. 1 (April 2009), pp. 19–38, and I. Capua and D. I. Aleksander, "Avian Influenza Infection in Birds: A Challenge and Opportunity for the Poultry Veterinarian," *Poultry Science*, vol. 88, no. 4 (April 1988), pp. 842–6. A highly pathogenic outbreak hit Texas in 2004. See Angela Pelzel et al., "Review of Highly Pathogenic Outbreak of Avian Influenza in Texas 2004," *Journal of the American Veterinary Medical Association*, vol. 228, no. 12 (June 15, 2006), pp. 1869–75.

83 Alexander and Brown, "History," pp. 19–38.

84 Nicita, *Avian Influenza and the Poultry Trade*, p. 1. As Nicita notes, "Because of high mortality rates, high rates of contagion, and the possibility of cross-species infection to mammals including humans, high pathogenic avian influenza is a major concern both to consumers and producers of poultry."

85 FAO, *Biosecurity for Highly Pathogenic Avian Influenza* (Rome: FAO, 2008), p. 2.

86 Nicita, *Avian Influenza and the Poultry Trade*, pp. 4–5.

87 Pew Commission on Industrial Farm Animal Production, *Putting Meat on the Table: Industrial Farm Production in America* (Philadelphia: Pew, 2008), p. 11.

88 Pew Commission, *Putting Meat on the Table*, p. 17.

89 Bryan Walsh, "Why Meat in China – and the U.S. – Has a Drug Problem," *Time*, February 12, 2013, at http://science.time.com/2013/02/12/why-meat-in-china-and-the-u-s-has-a-drug-problem.

4 Shit and Feathers

1 General Accounting Office (GAO), *Animal Agriculture Waste Management Practices*, Report GAO/RECD-99-205 (Washington, DC: GAO, July 1999), pp. 1–2.

2 Heribert Insam, Ingrid Franke-Whittle and Sabine Podmirseg, "Agricultural Waste Management in Europe, with an Emphasis on Anaerobic Digestion," *Journal of Integrated Field Science*, vol. 11 (2014), pp. 13–17, at www.researchgate.net/publication/265258151_Agricultural_Waste_Management_in_Europe_with_an_Emphasis_on_Anaerobic_Digestion.

3 L. Loyon, "Overview of Manure Treatment in France," *Waste Management*, vol. 61 (2017), pp. 516–20.

4 Chaohui Zheng et al. "Managing Manure from China's Pigs and Poultry: The Influence of Ecological Rationality," *Ambio*, vol. 43, no. 5 (2014), pp. 661–72, doi:10.1007/s13280-013-0438-y.

5 GAO, *Animal Agriculture Waste Management Practices*, p. 3.

6 John Cassius Moreki and Teto Keaikitse, "Poultry Waste Management Practices in Selected Poultry Operations around Gaborone, Botswana," *International Journal of Current Microbiology and Applied Sciences*, vol. 2, no. 7 (2013), pp. 240–8.

7 Pew Memorial Trust, *Chicken: Pollution and Industrial Poultry Production in America* (Washington, DC: Pew Environmental Group, July 2011), pp. 1–3, 10–11, 13.

8 Verel W. Benson and Thomas J. Witzig, *The Chicken Broiler Industry: Structure, Practices and Costs*, Agricultural Economic Report 381 (Washington, DC: USDA, 1971), pp. 18, 27–8. NB: The publication date on the brochure is 1971, but the information in it indicates a typographical error, and it should be 1976.

9 J. M. Roberts, "Combined Treatment of Poultry and Domestic Wastes," *Sewage and Industrial Wastes*, vol. 30, no. 9 (Sept. 1958), pp. 1186–7.

10 Roberts, "Combined Treatment of Poultry and Domestic Wastes," p. 1187.

11 Roberts, "Combined Treatment of Poultry and Domestic Wastes," p. 1189.

12 Pew, *Chicken: Pollution and Industrial Poultry Production*, p. 4.

13 Pew Environment Group, *Big Chicken: Pollution and Industrial Poultry Production in America* (July 2011), introduction and pp. 1–2, 4, 6–7.

14 James MacDonald and William McBride, *The Transformation of Livestock Agriculture: Scale, Efficiency and Risks*, Economic Research Service Information Bulletin 43 (Washington, DC: USDA, January 2009), p. 28.

15 Pew Commission on Industrial Farm Animal Production, *Putting Meat on the Table: Industrial Farm Production in America* (Philadelphia: Pew: 2008), p. 6.

16 Pew Commission, *Putting Meat on the Table*, p. 11.

17 Y. Lee et al., "Identifying Sources of Fecal Contamination in Streams Associated with Chicken Farms," paper presented at 108th

General Meeting of the American Society of Microbiology, Boston, Mass., June 1–5, 2008, at https://cfpub.epa.gov/si/si_public_record_report.cfm?Lab=NERL&dirEntryId=188431.

18 EPA, "Estimated Animal Agriculture Nitrogen and Phosphorus from Manure," *EPA*, January 30, 2019, at www.epa.gov/nutrient-policy-data/estimated-animal-agriculture-nitrogen-and-phosphorus-manure.

19 Pew Commission, *Putting Meat on the Table*, p. viii.

20 Environmental Integrity Project, "Chesapeake Bay and Factory Farms," at www.environmentalintegrity.org/what-we-do/chesapeake-bay-and-factory-farms.

21 Brian Lawson, "One Million Tons of Chicken Waste in Alabama Every Year. Where Does It All Go?" *AL.com*, March 22, 2015, at www.al.com/news/2015/03/alabama_farmers_have_to_deal_w.html.

22 EPA, *Managing Manure Nutrients at Concentrated Animal Feeding Operations*, EPA-821-B-04-009 (Washingon, DC: EPA, December 2004), pp. 1–4.

23 Pew, *Chicken: Pollution and Industrial Poultry Production*, pp. 52–3.

24 Pew, *Chicken: Pollution and Industrial Poultry Production*, pp. 17–18.

25 John Murawski, "Florence Kills 5,500 Pigs and 3.4 Million Chickens," *News and Observer*, September 18, 2018, at www.newsobserver.com/news/local/article218610365.html.

26 MacDonald and McBride, *The Transformation of Livestock Agriculture*, pp. 30–1.

27 Pew, *Big Chicken*, pp. 13–16.

28 Ecowatch, "Factory Farms Pollute the Environment and Poison Drinking Water," February 20, 2019, at www.ecowatch.com/factory-farms-drinking-water-pollution-2629508815.html.

29 Steven Sellers, "As Factory Farms Spread, So Do Toxic Tort Cases," *Bloomberg*, May 2, 2017, at https://news.bloombergenvironment.com/environment-and-energy/as-factory-forms-spread-so-do-toxic-tort-cases.

30 Neil M. Dubrovsky and Pixie A. Hamilton, *Nutrients in the Nation's Streams and Groundwater: National Findings and Implications*, Fact Sheet 2010-3078 (Reston, Va.: US Geological Survey, 2010).

31 I rely heavily here on Lawson, "One Million Tons of Chicken Waste," for much of the analysis in the following paragraphs.
32 Lawson, "One Million Tons of Chicken Waste."
33 Lawson, "One Million Tons of Chicken Waste."
34 Paul L. Hollis Farm Press, "From Nov. 15 to Feb. 15: Alabama Enacts Poultry Litter Ban," January 2, 2002, at www.farmprogress.com/nov-15-feb-15-alabama-enacts-poultry-litter-ban.
35 Doug Gurian-Sherman, *CAFOs Uncovered: The Untold Costs of Concentrated Animal Feed Operations* (Cambridge, Mass.: Union of Concerned Scientists, April 2008), p. 2.
36 Carl Furiness et al., "Forests as an Alternative for Poultry Manure Application," AG-739, NC State Extension, March 25, 2019, at https://content.ces.ncsu.edu/forests-as-an-alternative-for-poultry-manure-application.
37 John Blake et al., "Poultry Carcass Disposal Options for Routine and Catastrophic Mortality," *Cast Issue Paper*, no. 40 (October 2008), pp. 1–2.
38 D. J. Shafer et al., "Chemical Preservation of Whole Broiler Carcasses Utilizing Alkaline Hydroxides," *Poultry Science*, vol. 79, no. 11 (2000), pp. 1517–23.
39 Blake et al., "Poultry Carcass Disposal," pp. 1–2.
40 Blake et al., "Poultry Carcass Disposal," pp. 1–2.
41 "Industrial Feather Waste Valorisation for Sustainable KeRatin-based Materials," at www.karma2020.eu.
42 R. J. Buhr et al., "Influence of Flooring Type During Transport and Holding on Bacteria Recovery from Broiler Carcass Rinses Before and After Defeathering," *Poultry Science*, vol. 79, no. 3 (2000), pp. 436–41.
43 Robert Goodland and Jeff Anhang, "Livestock and Climate Change," *World Watch* (November/December 2009), pp. 10–19.
44 FAO, *Livestock's Long Shadow: Environmental Issues and Options* (Rome: FAO, 2006); Giampiero Grossi et al., "Livestock and Climate Change: Impact of Livestock on Climate and Mitigation Strategies," *Animal Frontiers*, vol. 9, no. 1 (January 2019), pp. 69–76; and Goodland and Anhang, "Livestock and Climate Change."
45 V. de Sy et al., "Land Use Patterns and Related Carbon Losses Following Deforestation in South America," *Environmental*

Research Letters, vol. 10, no. 12 (2015), at www.cifor.org/library/5892; and Global Forest Coalition, "What's at Steak? The Real Cost of Meat," December 5, 2016, at https://globalforestcoalition.org/whats-steak-real-cost-meat.

46 Dario Caro et al., "CH4 and N_2O Emissions Embodied in International Trade of Meat," *Environmental Research Letters*, vol. 9 (2014), p. 1.

47 Caro et al., "CH4 and N_2O Emissions," pp. 4–5

48 Caro et al., "CH4 and N_2O Emissions," pp. 7–8.

49 Gurian-Sherman, *CAFOs Uncovered*, p. 2.

50 Gurian-Sherman, *CAFOs Uncovered*, p. 3.

51 Gurian-Sherman, *CAFOs Uncovered*, pp. 3–4.

52 Gurian-Sherman, *CAFOs Uncovered*, p. 4.

53 Pew, *Chicken: Pollution and Industrial Poultry Production*, pp. 1–3, 10–11, 13.

54 Helena Bottemiller, "FDA Urged to Ban Poultry Waste in Feed," *Food Safety News*, March 2, 2009, at www.foodsafetynews.com/2009/11/petition-seeks-to-ban-feeding-chicken-feces-to-cattle/#.UmWjo7vLiyM.

55 Pew Commission, *Putting Meat on the Table*, p. 23.

5 Pecking and Protest

1 See, for example, Humane Society of the United States, "Smithfield Foods (2012 Webby Award Winner)," December 15, 2010, at www.youtube.com/watch?v=L_vqIGTKuQE; Tompkins Conservation, "CAFO, The Book – Video Trailer," September 1, 2010, at www.youtube.com/watch?v=AOVn7_drjkE; Compassionate Consumers, "Wegmans Cruelty," 2005, at https://topdocumentaryfilms.com/wegmans-cruelty.

2 Upton Sinclair, *The Jungle* (New York: Signet, 1960), pp. 100–2.

3 Dovainoniu paukstynas, "About Us," at www.visciukai.lt/10369/about-us.html, and Vilniaus paukštynas and Kaisiadorys, "KG Group plans: more efficient and environmentally friendly poultry processing line," at www.paukstynas.eu/en/news/kg-group-plans-more-efficient-and-environmentally-friendly-poultry-processing-line.

4 Among the scores of articles on the disease, see Peter Smith, "The

Epidemics of Bovine Spongiform Encephalopathy and Variant Creutzfeldt–Jakob Disease: Current Status and Future Prospects," *Bulletin of the World Health Organization*, vol. 81 (2003), pp. 123–30.

5 UK World Animal Protection, *Pecking Order: Fast Food Giants are Failing Chickens* (2018) at www.worldanimalprotection.org/sites/default/files/int_files/the_pecking_order_full_report.pdf?_ga=2.187104289.847455312.1552505801-1289079408.1552505801.

6 Compassion in World Farming, "About Chickens Farmed for Meat," at www.ciwf.org.uk/farm-animals/chickens/meat-chickens.

7 Food Safety News, "Nation's Oldest 'Ag-Gag' Law Challenged by Animal Legal Activists," at www.foodsafetynews.com/2018/12/nations-oldest-ag-gag-law-challenged-by-animal-legal-activists/ December 6, 2018.

8 ALEC, "The Animal and Ecological Terrorism Act (AETA)," www.alec.org/model-policy/the-animal-and-ecological-terrorism-act-aeta. See also "Pre-emption of Local Agriculture Laws Act," at www.alec.org/issue/agriculture.

9 Richard Oppel, Jr., "Taping of Farm Cruelty Is Becoming the Crime," *New York Times*, April 6, 2013, at www.nytimes.com/2013/04/07/us/taping-of-farm-cruelty-is-becoming-the-crime.html.

10 The Bathroom Bill led to a national boycott of various events in North Carolina, and a loss of at least $600 million in state revenues. It remains unclear whether the same impotent state inspectors who glance away from environmental and animal cruelty violations would have been used to verify genitalia at the door of every bathroom.

11 AP (Associated Press), "Court Restores Lawsuit against North Carolina 'Ag-gag' Law," June 5, 2018, at www.apnews.com/587c8986377840319ded7dc455869cdd; Eric Frazier, "N.C.'s Boneheaded 'Ag-Gag' Law Protects Corporate Wrongdoing from Exposure," August 10, 2018, at www.charlotteobserver.com/opinion/opn-columns-blogs/o-pinion/article53141640.html.

12 Editorial Board, "No More Exposés in North Carolina," *New York Times*, February 1, 2016, at www.nytimes.com/2016/02/01/opinion/no-more-exposes-in-north-carolina.html.

13 Greenpeace, "Corporate Bullies Can't Silence the Resistance," at https://web.archive.org.

14 See Court of Appeal, First District, Division 1, California. *Physicians*

Committee for Responsible Medicine, Plaintiff and Appellant, v. *Tyson Foods, Inc., Defendant and Respondent*. No. A103835. Decided: June 1, 2004 at https://caselaw.findlaw.com/ca-court-of-appeal/1445542.html.

15 Editors, "Iowa's Ag-Gag Law Has Lingered Long Enough," *Des Moines Register*, October 26, 2017, at https://eu.desmoinesregister.com/story/opinion/editorials/2017/10/25/iowa-ag-gag-law-confinement-fraud-puppy-mill/797118001.

16 Matt Reynolds, "Iowa Must Defend Controversial 'Ag-Gag' Law," March 1, 2018, www.courthousenews.com/iowa-must-defend-controversial-ag-gag-law.

17 Donnelle Eller, "Lawsuit Saying Iowa's Ag Gag Law Is Unconstitutional Moves Forward," February 28, 2018, at https://eu.desmoinesregister.com/story/money/agriculture/2018/02/28/lawsuit-iowa-ag-gag-law-unconstitutional-moves-forward/380792002.

18 Bill Chappell, "Judge Overturns Utah's 'Ag-Gag' Ban on Undercover Filming at Farms," July 8, 2017, at www.npr.org/sections/thetwo-way/2017/07/08/536186914/judge-overturns-utahs-ag-gag-ban-on-undercover-filming-at-farms.

19 Paul Crenshaw, "Thinking of Chickens," *Southwest Review*, vol. 97, no. 4 (2012), pp. 562–3.

20 Brian Bienkowski, "Hog Waste and the End of the Family Farm," Nov. 13, 2017, at www.ehn.org/hog-waste-2504936054.html?rebelltitem=1#rebelltitem1.

21 Natalia Dannenberg, "Protesters in Berlin Call For an End to Factory Farming," *DeutscheWelle*, January 22, 2011 at www.dw.com/en/protesters-in-berlin-call-for-an-end-to-factory-farming/a-14780207-1.

22 "The Leading Agricultural Country in the EU," www.terresdeurope.net/en/france-agriculture-leading-country-europe.asp.

23 CIWF France, "Poulet de Chair" (2019), at www.ciwf.fr/animaux-de-ferme/poulets-de-chair.

24 Compassion in World Farming, "Paris Demo: 'No to Factory Farming,'" March 6, 2013, at www.ciwf.org.uk/news/2013/03/paris-demo-no-to-factory-farming.

25 France24, "Factory Farming Is on the Rise in France, Union Says,"

February 20, 2015, at www.france24.com/en/20150220-factory-farms-rise-france-agricultural-union.

26 Novissen, "Pourquoi NOVISSEN? Pourquoi engager cette lutte depuis six ans?" at http://novissen.com/pourquoi-novissen.

27 Novissen, "NOs VIllages Se Soucient de leur ENvironnement," http://novissen.com.

28 Sofiprotéol, "Sofiprotéol, partenaire stratégique de l'agro-industrie et de l'agroalimentaire," at www.sofiproteol.com.

29 Confédération paysanne, "Carte de l'industrialisation de l'agriculture: une dérive destructrice pour les paysans," February 19, 2015, at www.confederationpaysanne.fr/actu.php?id=3347&PHPSESSID=n79v9ds53f7sq12fjiv3vdf8i2.

30 "Angry French Farmers Hold Tractor Protest in Paris," *The Guardian*, September 3, 2015, at www.theguardian.com/world/2015/sep/03/angry-french-farmers-hold-tractor-protest-in-paris.

31 L214, Film "Stop à l'élevage intensif des poulets," "En France, 8 poulets sur 10 sont détenus en élevage intensif," at www.l214.com/poulets/800-millions.

32 L214, "L'impact de l'élevage des poulets sur la santé et l'environnement," www.l214.com/impact-de-elevage-des-poulets-sur-la-sante-et-environnement.

33 "Nouveau scandale alimentaire dans un élevage de 200 000 poules," *La Croix*, May 25, 2016, at www.la-croix.com/Economie/France/Nouveau-scandale-alimentaire-dans-elevage-200-000-poules-2016-05-25-1200762771. See the film clip, L214, "Vie de Misère Pour des Poulets en Auvergne" at www.l214.com/enquetes/2019/elevage-made-in-france/poulets-auvergne.

34 L214, "L'impact de l'élevage des poulets."

35 Tass, "Putin Signs Bill to Ban All Forms of Cruelty to Animals," December 28, 2018, at http://tass.com/society/1038276; and "Путин подписал закон об ответственном обращении с животными," *Novaia Gazeta*, December 28, 2018, at www.novayagazeta.ru/news/2018/12/28/148001-putin-podpisal-zakon-ob-otvetstvennom-obraschenii-s-zhivotnymi.

36 Vladimir Tarlo, "Россия-пыточная для бездомных животных," https://democrator.ru/petition/rossiya-pytochnaya-dlya-bezdomnyh-zhivotnyh.

37 Tarlo, "Россия-пыточная."

38 Tarlo, "Россия-пыточная." See also "Правда о суете вокруг Закона о животных," May 19, 2017, at https://zoohumanism.com/2017/05/19/ правда-о-суете-вокруг-закона-о-животны.

39 ESDAW (European Society of Dog and Animal Welfare), "Society and Animal Welfare – Russia," at www.esdaw.eu/society-and-animal-welfare---russia.html.

40 Amnesty International, "Russia: A Year On, Putin's 'Foreign Agents Law' Choking Freedom," November 20, 2013, at www.amnesty.org/en/news/russia-year-putin-s-foreign-agents-law-choking-freedom-2013-11-20.

41 LAPA, "Helping Animals in Russia," at www.lapauk.org/about.

42 ESDAW, "Society and Animal Welfare." See also C. J. Chivers, "A Brutal Sport Is Having Its Day Again in Russia," *New York Times*, February 9, 2007, at www.nytimes.com/2007/02/09/world/europe/09dogfight.html?_r=1&ei=5090&en=c2d1b84093319843&ex=1328677200&partner=rssuserland&emc=rss&pagewanted=all.

43 A Communist Party Senator, Sergei Kalashnikov, warned that a bill outlawing cruelty to dogs was a "slippery slope" to LGBT rights, a position that truly drives home the point that people who are cruel to animals will be cruel to each other and seek to justify discrimination again others based on claims of "other-ness": "Russian Senator Compares Animal Cruelty Law to Gay Rights," *Moscow News*, December 26, 2016, at https://themoscowtimes.com/news/russian-senator-compares-animal-cruelty-law-to-gay-rights-60056.

44 Peter Stevens et al., *Review of Animal Welfare Legislation in the Beef, Pork, and Poultry Industries* (Rome: FAO, 2014), pp. 22–3.

45 FAO, *Review of Animal Welfare Legislation*.

46 Nanda Ursinus et al., "General Overview of Animal Welfare in a Global Perspective," Report 240, at https://core.ac.uk/download/pdf/29251008.pdf.

47 *Poultry World News*, Dec, 8, 2014 at www.poultryworld.net/Meat/Articles/2014/12/Only-four-countries-receive-highest-animal-welfare-rating-1661627W.

48 Stephanie Strom, "After Suspending Supplier, Chipotle Takes Pork Off Menu in 600 Stores," *New York Times*, January 14, 2015, at www.nytimes.com/2015/01/15/business/after-suspending-supplier-

chipotle-takes-pork-off-menu-in-600-stores.html; see "Day After Day We are Committed," at www.chipotle.com/food-with-integ rity?format=aspx#.

49 Christopher Doering, "Smithfield Urges Farmers to End Use of Gestation Crates," *USA Today*, January 7, 2014, at https://eu.usa today.com/story/news/nation/2014/01/07/hog-crates-ban/4362353.

50 Kari Hamerschlag and Sasha Stashwick, *Chain Reaction* (n.p., September 2016), p. 4.

51 Hamerschlag and Stashwick, *Chain Reaction*, p. 5.

52 Hamerschlag and Stashwick, *Chain Reaction*, p. 5.

53 Pew Charitable Trusts, *The Business of Broilers: Hidden Costs of Putting a Chicken on Every Grill* (Philadelphia, Pa.: Pew, December 2013), p. 16.

54 Pew, *The Business of Broilers*, pp. 1–2.

55 Eric Schlosser, *Fast Food Nation* (New York: Houghton Mifflin, 2001).

56 David Griffith, "Consequences of Immigration Reform for Low-Wage Workers in the Southeastern U.S.: The Case of the Poultry Industry," *Urban Anthropology and Studies of Cultural Systems and World Economic Development*, vol. 19, no. 1/2 (Spring–Summer 1990), pp. 156–63.

57 Ronny Rojas, Almudena Toral and Antonio Cucho, "Fear in the Chicken Capital of the United States," Univision, November 1, 2017, at www.univision.com/univision-news/immigration/in-the-countrys-chicken-processing-capital-undocumented-immigrants-live-in-fear.

58 Mehdi Gherdane, "Yvelines: l'arrivée de 40 000 poules inquiète les habitants de Gressey," *Le Parisien*, May 3, 2018, at www.leparisien.fr/yvelines-78/yvelines-l-arrivee-de-40-000-poules-inquiete-les-habitants-de-gressey-03-05-2018-7696230.php.

59 Robyn Alders and Robert Pym, "Village Poultry: Still Important to Millions, Eight Thousand Years after Domestication," *World's Poultry Science Journal*, vol. 65, no. 2 (2009), pp. 183–4.

60 Alders and Pym, "Village Poultry," pp. 183–4.

61 Jay Istvanffy, *A Guide to Aboriginal Harvesting Rights* (Vancouver: Legal Services Society, 2011).

62 See, for example, Legal Services Society, *A Guide to Aboriginal*

Hunting Rights (Legal Services Society: British Columbia, December 2017) at https://pubsdb.lss.bc.ca/resources/pdfs/pubs/A-Guide-to-Aboriginal-Harvesting-Rights-eng.pdf.

63 Stanislaus J. Dundon, "Agricultural Ethics and Multifunctionality Are Unavoidable," *Plant Physiology*, vol. 133, no. 2 (October 2003), pp. 427–37.

64 Dundon, "Agricultural Ethics and Multifunctionality."

65 Pew Commission on Industrial Farm Animal Production, *Putting Meat on the Table: Industrial Farm Production in America* (Philadelphia: Pew, 2008), pp. 24–5.

66 Pew Commission, *Putting Meat on the Table*, p. vii.

6 Drumsticks

1 International Agency for Research on Cancer, *Red and Processed Meat*, vol. 114 (IARC: Lyon, France, 2018).

2 Michael Wines, "Russia's Latest Export: Bad Jokes about US Chickens," *New York Times*, March 2, 2002, at www.nytimes.com/2002/03/02/world/russia-s-latest-export-bad-jokes-about-us-chickens.html; Michael Schwirtz, "Russia Seeks to Cleanse Its Palate of U.S. Chicken," *New York Times*, January 19, 2010, at www.nytimes.com/2010/01/20/world/europe/20russia.html?_r=0; Ian Traynor, "Bush's Legs Walk All Over Russians," *The Guardian*, March 9, 2002, at www.theguardian.com/world/2002/mar/09/russia.iantraynor.

3 George Watts, President, National Chicken Council, "The U.S. Poultry Industry: Competing in the International Market and Meeting Increasing Consumer Demand in the U.S.," March 18, 2005, National Chicken Council, at www.nationalchickencouncil.org/the-u-s-poultry-industry-competing-in-the-international-market-and-meeting-increasing-consumer-demand-in-the-u-s.

4 US International Trade Commission, *Industry and Trade Summary: Poultry*, Publication 2520 (AG-6) (Washington, DC: US ITC, June 1992), p. 8.

5 USDA, Foreign Agricultural Service (FAS), *Livestock and Poultry: World Markets and Trade*, Circular Series DL&P 2-01 (Washington, DC: DOA, October 2001), pp. 1–2.

6 Marilia Ferreira et al., *The Saga of the Brazilian Poultry Industry:*

How Brazil Has Become the World's Largest Exporter of Chicken Meat (São Paulo: UBABEF [União Brasileira de Avicultura, the Brazilian Poultry Union], 2011), p. 6.

7 Ferreira et al., *Saga*, pp. 15–16, 19, 22, 24.

8 Ferreira et al., *Saga*, p. 26.

9 Ferreira et al., *Saga*, pp. 28, 30, 32, 34.

10 Ferreira et al., *Saga*, pp. 42, 44, 46.

11 Ferreira et al., *Saga*, pp. 52–4.

12 Brazilian Chicken Producers and Exporters Association (ABEF), *Brazilian Chicken: Quality, Sustainability and Leadership, Brazil Now* (São Paulo: ABEF, 2008), p. 22.

13 Ferreira et al., *Saga*, p. 56.

14 ABPA (Brazilian Animal Protein Association), "ABPA History," at www.brazilianchicken.com.br/en/abpa-apexbrazil/abpa.

15 ABPA, "Background," at www.brazilianchicken.com.br/en/poultry-industry/background.

16 ABPA, "The Sustainability and Quality of the Sector," at www.brazilianchicken.com.br/en/poultry-industry/sustainability.

17 Ferreira et al., *Saga*, pp. 56, 61.

18 Ferreira et al., *Saga*, p. 76.

19 Greenpeace, "'Soya King' wins Golden Chainsaw Award," June 19, 2005, at www.greenpeace.org/archive-international/en/news/features/soya-king-wins-chainsaw.

20 Robert Soutar and Cristina Veiga, "Brazil Beef Scandal Highlights Dangers of Industrial Livestock Farming," *China Dialogue*, March 27, 2017, at www.chinadialogue.net/article/show/single/en/9698-Brazil-beef-scandal-highlights-dangers-of-industrial-livestock-farming; Bloomberg, "Frozen Beef Stranded at Sea as China Shuts Out Brazil's Meat," March 21, 2017, at www.bloomberg.com/news/articles/2017-03-22/frozen-beef-stranded-at-sea-after-china-shuts-out-brazil-s-meat; "Reestablished China–Brazil Beef Trade Means More Than Cheap Churrasco," May 28, 2015, at https://dialogochino.net/2489-reestablished-china-brazil-beef-trade-means-more-than-cheap-churrasco.

21 ABEF, *Brazilian Chicken*, p. 14.

22 US International Trade Commission, *Industry and Trade Summary*, p. 9.

23 Manuela Andreoni, "China Made Brazil a Global Agricultural Powerhouse. But Who Benefits?" *DialogoChino*, January 7, 2019, at https://dialogochino.net/19746-china-made-brazil-a-global-agricultural-powerhouse-but-who-benefits.
24 US International Trade Commission, *Industry and Trade Summary*, p. 9.
25 Chendong Pi, *Fair or Fowl? Industrialization of Poultry Production in China* (Minneapolis: Institute for Agriculture and Trade Policy, February 2014).
26 USDA, FAS, *Livestock and Poultry*, p. 24.
27 Pi, *Fair or Fowl?*
28 Rong-fa Guan et al., "Meat Quality Traits of Four Chinese Indigenous Chicken Breeds and One Commercial Broiler Stock," *Journal of Zhejiang University Science B*, vol. 14, no. 10 (October 2013), pp. 896–902, at www.ncbi.nlm.nih.gov/pmc/articles/PMC3796641.
29 Pi, *Fair or Fowl?*
30 Reuters, "China's Wen's Overhauls Poultry Business after Bird Flu Toll," October 30, 2017, at www.reuters.com/article/china-guangdong-wens-results/chinas-wens-overhauls-poultry-business-after-bird-flu-toll-idUSL4N1N5215.
31 Ke Bingsheng and Han Yijun, "Poultry Sector in China: Structural Changes During the Past Decade and Future Trends," *Poultry in the Twentieth Century* (Research Center for Rural Economy [RCRE], Ministry of Agriculture, China, 2008[?]), pp. 1, 30, at https://pdfs.semanticscholar.org/0e3c/bfc507a8c9fa37f1c773d57158fb9b1fc354.pdf.
32 Fang Xiao and Virginia Wu, "Chinese KFC Chicken Supplier in Feed Scandal," *EpochTimes*, November 28, 2012, at www.theepochtimes.com/chinese-kfc-chicken-supplier-in-feed-scandal_1479942.html. The original report (in Chinese) is at www.ce.cn/cysc/sp/info/201211/27/t20121127_21291146.shtml.
33 USDA, FAS, "China: Chinese Chicken Meat Production Continues to Recover," September 12, 2018, at www.fas.usda.gov/data/china-chinese-chicken-meat-production-continues-recover.
34 Dominique Patton, "China's Poultry Giant Goes to Lab in Quest for Meatier Chickens," June 12, 2018, at www.reuters.com/article/

us-china-poultry-guangdong-wens-focus/chinas-poultry-giant-goes-to-lab-in-quest-for-meatier-chickens-idUSKBN1J831C.

35 Andy Uhler, "Chinese Tariffs Make it Tough for U.S. Chicken Feet," May 10, 2016, at www.marketplace.org/2016/05/10/world/chinese-tariffs-make-it-tough-export-chicken-there.

36 K. William Watson, "Antidumping Fowls Out: U.S. – South Africa Chicken Dispute Highlights the Need for Global Reform," CATO Institute, October 19, 2015, at www.cato.org/publications/free-trade-bulletin/antidumping-fowls-out-us-south-africa-chicken-dispute-highlights.

37 G. Pascal Zachary, "Cheap Chickens: Feeding Africa's Poor," *World Policy Journal*, vol. 21, no. 2 (Summer 2004), pp. 47–52.

38 Zachary, "Cheap Chickens," pp. 47–9.

39 Olivier Roger, "Le poulet européen fâche l'Afrique," RFI, March 6, 2018, at www.rfi.fr/emission/20180306-poulet-europe-fache-afrique-producteurs-volailles-exportation.

40 See Kwaw Andam, Channing Arndt and Faaiqa Hartley, *Eggs Before Chickens? Assessing Africa's Livestock Revolution with an Example from Ghana*, IFPRI Discussion Paper 1, 2017, at www.ifpri.org/publication/eggs-chickens-assessing-africas-livestock-revolution-example-ghana.

41 Roger, "Le poulet européen fâche l'Afrique."

42 AVEC, "Our Vision," www.avec-poultry.eu/what-we-do/our-vision.

43 Dmitry Prikhodko and Albert Davleyev, *Russian Federation: Meat Sector Review* (Rome: FAO/ERBD [European Bank for Reconstruction and Development], 2014), p. xiv.

44 US International Trade Commission, *Industry and Trade Summary*, p. 9.

45 USDA, FAS, *Livestock and Poultry*, p. 27.

46 Svetlana Mentiukova, "Юрий Лужков запретил 'ножки Буша'," *Kommersant*, May 20, 2002, at www.kommersant.ru/doc/322841?query=%D0%BA%D1%83%D1%80%D1%8F%D1%82%D0%B8%D0%BD%D0%B0.

47 Elena Evstigneeva, "Птицеводы взялись за перо," *Vedomosti*, December 25, 2001, at www.vedomosti.ru/newspaper/articles/2001/12/25/pticevody-vzyalis-za-pero.

48 Mikhail Overchenko, "Американцы объелись дешевым мясом" *Vedomosti*, April 26, 2002, at www.vedomosti.ru/newspaper/articles/2002/04/26/amerikancy-obelis-deshevym-myasom.

49 Sof'ia Korepanova, "Путин пугает США птицей," *Vedomosti*, August 29, 2008, at www.vedomosti.ru/newspaper/articles/2008/08/29/putin-pugaet-ssha-pticej.

50 N.a., "Хлорированную курятину приняли в Таможенный союз," *Kommersant*, September 14, 2010, at www.kommersant.ru/doc/1504140?query=%D0%BA%D1%83%D1%80%D1%8F%D1%82%D0%B8%D0%BD%D0%B0.

51 Anfisa Voronina and Natal'ia Kostenko, "Куриная война с Америкой окончена," *Vedomosti*, June 28, 2010, at www.vedomosti.ru/business/articles/2010/06/28/kurinaya-vojna-s-amerikoj-okonchena.

52 Sergei Sysoikin, "Курица нон-грата," *Kommersant*, January 12, 2012, at www.kommersant.ru/doc/1848909?query=%D0%BA%D1%83%D1%80%D1%8F%D1%82%D0%B8%D0%BD%D0%B0.

53 Dmitri Ladygin, "'Черкизово' поглотит 'Куриное царство," *Kommersant*, April 18, 2007, at www.kommersant.ru/doc/760076?query=%D0%BA%D1%83%D1%80%D1%8F%D1%82%D0%B8%D0%BD%D0%B0.

54 Natal'ia Starostina and Iurii L'vov, "Куриный оскал," *Kommersant*, August 20, 2007, at www.kommersant.ru/doc/796071?query=%D0%BA%D1%83%D1%80%D1%8F%D1%82%D0%B8%D0%BD%D0%B0.

55 Watts, "The U.S. Poultry Industry."

56 USDA, FAS, *Livestock and Poultry: World Markets and Trade* (USDA, FAS, October 11, 2018), p. 22.

57 Sipke Joost Hiemstra and Jan Ten Napel, *Study of the Impact of Genetic Selection on the Welfare of Chickens Bred and Kept for Meat Production*, Final Report, IBF International Consulting, January 2013, p. 9.

58 US Government Census Bureau, "National Poultry Day," March 19, 2018, at www.census.gov/newsroom/stories/2018/poultry.html.

59 USDA FAS, *Livestock and Poultry*, October 11, 2018, p. 22.

60 Watts, "The U.S. Poultry Industry."

Epilogue: Broiler Chernobyl

1 Viktor Pelevin, "Hermit and Sixfinger," trans. Serge Winitzki and Sergey Bratus, copyright 1996, at сайт творчества Виктор Пелевин, http://pelevin.nov.ru/pov/en-hermit/1.html.

2 Alfred Crosby, *Ecological Imperialism* (Cambridge University Press, 1986).

3 Carolyn Merchant, *The Death of Nature* (San Francisco: Harper Collins, 1990), pp. 164–252.

4 From Raj Patel and Jason W. Moore, *A History of the World in Seven Cheap Things: A Guide to Capitalism, Nature, and the Future of the Planet* (Berkeley: University of California Press, 2018), as excerpted in Patel and Moore, "How the Chicken Nugget Became the True Symbol of our Era," *The Guardian*, May 8, 2016, at www.theguardian.com/news/2018/may/08/how-the-chicken-nugget-became-the-true-symbol-of-our-era.

5 Pew Environment Group, *Big Chicken: Pollution and Industrial Poultry Production in America* (July 2011), p. 2. See also Compassion in World Farming, *The Life of: Broiler Chickens*, at www.ciwf.org.uk/media/5235306/The-life-of-Broiler-chickens.pdf.

6 For collections of chicken sayings, phrases and so on, see www.backyardchickens.com/articles/common-chicken-sayings-idioms-other-funny-things-we-say.47662 and http://natureessays.blogspot.com/2008/12/of-chickens-and-cliches.html.

Index